JOHN ARTHUR

The Biology of the Human Mind

How Brain, Hormones, and Genetics Shape Our Thoughts, Emotions, and Behavior

First edition

This book was professionally typeset on Reedsy.
Find out more at reedsy.com

Contents

Foreword

The human brain is the most complex structure in the known universe. Within the roughly three-pound mass inside your skull reside approximately 86 billion neurons, each forming thousands of connections with other neurons, creating a network of staggering intricacy. This biological organ generates every thought you have ever had, every emotion you have ever felt, every memory you cherish, and every dream that visits you in sleep. It is simultaneously the most familiar aspect of your existence you experience its products every waking moment and profoundly mysterious.

For most of human history, the relationship between brain and mind remained obscure. Ancient philosophers debated whether the mind was separate from the body, a spiritual entity temporarily housed in physical form. Even as scientific understanding advanced, the brain was often viewed as too complex to understand, its workings forever beyond human comprehension. The remarkable progress of neuroscience over the past century has shattered these limitations, revealing that the mind is not separate from the brain but emerges from it, and that the biological processes creating mental life can be studied, understood, and sometimes modified to reduce suffering and enhance human capacities.

This book invites you on a journey through the biological foundations of human psychology. You will discover how neurons communicate through electrical and chemical signals, how different brain regions support different mental functions, how hormones influence mood and behavior, how memories are formed and stored, how emotions emerge from coordinated activity across brain structures, and how understanding these processes illuminates both normal functioning and what goes wrong in mental illness.

The implications extend far beyond the laboratory. Understanding the biology of psychology has transformed how we approach mental health, education, criminal justice, and human potential. It has reduced stigma around mental illness by revealing these conditions as brain disorders rather than character flaws. It has led to effective treatments for depression, anxiety, and other conditions that cause immense suffering. It has shown that intelligence is not fixed but can be cultivated through appropriate challenges. It has revealed the brain's remarkable capacity for plasticity, reorganizing throughout life in response to experience.

Yet biological psychology does not reduce human experience to mere mechanism. Rather, it reveals the extraordinary sophistication of the biological processes that create the richness of mental life. It shows how genes and environment intertwine in complex ways, how early experiences shape neural development while adult brains retain capacity for change, and how the same biological foundations support vast diversity in human thought, emotion, and behavior.

This book is written for anyone curious about the biological basis of who we are. Whether you are a student beginning to explore psychology and neuroscience, an educator seeking to understand how the brain learns, a healthcare professional wanting deeper insight into mental health, or simply someone fascinated by the question of how three pounds of biological tissue creates a human mind, you will find accessible explanations grounded in rigorous science.

The journey ahead will take you from the molecular mechanisms of synaptic transmission to the systems-level organization of memory and emotion, from the genetic influences on personality to the environmental factors that shape development, from the neural basis of consciousness to the biology of mental illness and its treatment. Along the way, you will encounter remarkable case studies, groundbreaking experiments, and cutting-edge research that continues to expand our understanding.

Understanding biological psychology changes how you see yourself and others. It reveals that your thoughts and feelings, far from being abstract and immaterial, reflect physical processes in your brain that can be measured and understood. It shows that mental health and mental illness exist on continua rather than being discrete categories, and that seeking help for psychological suffering is no different from seeking help for physical illness. It demonstrates that learning and growth remain possible throughout life because the brain retains capacity for change.

As you read, you will be using your brain to understand how your brain works—a uniquely reflexive endeavor that gives neuroscience special character among the sciences. The neurons firing as you process these words, the memory systems encoding information you find interesting, the attention networks focusing on the page, and the emotional responses you experience are all products of the biological processes described in these chapters.

Welcome to the fascinating world of biological psychology, where science illuminates the most intimate aspects of human experience and where understanding the brain provides keys to living more fully, thinking more clearly, and approaching mental health with both compassion and evidence-based interventions.

Preface

This book emerged from a conviction that understanding the biological foundations of psychology is essential for anyone seeking to understand human nature, improve mental health, or unlock human potential. The past century has witnessed extraordinary advances in our knowledge of how the brain creates mind, yet this knowledge often remains confined to scientific journals and academic textbooks. My goal in writing this book has been to make this crucial knowledge accessible to general readers while maintaining scientific rigor and avoiding oversimplification.

The inspiration for this project came from years of observing how biological understanding transforms people's relationships with their own minds and mental health. Students encountering neuroscience for the first time often describe revelation and relief revelation at discovering the elegance and complexity of neural mechanisms, and relief at understanding that psychological struggles have biological underpinnings rather than reflecting personal inadequacy. People suffering from depression, anxiety, or other mental health conditions often find validation and hope in learning that their symptoms reflect measurable changes in brain chemistry and function that can be addressed through evidence-based treatments.

Yet I have also witnessed misunderstandings and misapplications of biological psychology. Oversimplified accounts of "chemical imbalances" obscure the complexity of mental illness. Claims about "left-brained" versus "right-brained" thinking persist despite being contradicted by evidence. Neuromyths about how much of the brain we use or critical periods that rigidly determine development proliferate in popular culture. These distortions can mislead people about their own capabilities and the nature of psychological problems.

This book seeks to present biological psychology accurately while remaining accessible, to convey both the power and the limits of current knowledge, and to explore implications for individuals and society while acknowledging ethical complexities. Each chapter builds on previous chapters while also standing alone, allowing readers to explore topics in any order based on interest.

The scope is broad by necessity, covering topics from molecular neuroscience to social psychology, from brain anatomy to mental illness, from development to aging. This breadth reflects the reality that understanding psychology requires integrating multiple levels of analysis, from genes to cultures, and that the most interesting questions often lie at the intersections of traditionally separate domains.

Throughout the book, I have tried to ground abstract concepts in concrete examples and to illustrate principles with real cases and research studies. Case studies of patients with brain damage reveal which brain regions support which functions. Research on London taxi drivers demonstrates neuroplasticity in action. Studies of stress hormones connect everyday experiences to biological processes. These examples make biological psychology come alive, showing how scientific findings illuminate real human experiences.

I have also aimed to present balanced perspectives on controversial topics including the nature-nurture debate, genetic influences on behavior, the biological basis of mental illness, and the potential for cognitive enhancement. These issues involve both scientific questions about how things work and value questions about how knowledge should be used. I have tried to present the science clearly while acknowledging that applications require ethical deliberation and societal choices that science alone cannot answer.

The organization progresses from foundational topics through specific psychological domains to applications and future directions. Early chapters establish basics of brain structure, neural communication, and nervous system organization. Middle chapters explore emotions, stress, social bonding, memory, learning, and intelligence. Later chapters examine sleep and mental health. The final chapter looks toward the future, considering emerging technologies, unresolved questions, and ethical challenges.

Readers with different backgrounds and interests may approach the book differently. Students may read sequentially to build comprehensive understanding. Healthcare professionals might focus on chapters addressing mental health and treatment. Educators could emphasize chapters on learning, memory, and intelligence. Readers curious about specific topics can dive directly into relevant chapters. I have tried to make each chapter comprehensible on its own while providing cross-references to related material elsewhere.

Writing this book has deepened my own appreciation for the remarkable organ we carry in our skulls and for the scientists whose patient work has illuminated its mysteries. From the pioneers who first mapped brain functions through studies of brain-damaged patients, to the researchers who discovered neurotransmitters and decoded neural signals, to the contemporary scientists using cutting-edge technologies to observe living brains in unprecedented detail, the cumulative effort of thousands of researchers has created our current understanding.

This knowledge is not complete profound mysteries remain about consciousness, memory, and how neural activity creates subjective experience. But what has been learned provides powerful tools for understanding ourselves and for addressing problems that cause immense suffering. My hope is that this book makes this knowledge accessible and useful to readers from all backgrounds, contributing to a world where biological understanding informs compassionate and effective approaches to mental health and human development.

The brain you are using to read these words is a product of millions of years of evolution, shaped by your unique genetic inheritance and your life experiences. It has changed since you began reading this preface and will continue changing throughout your life. Understanding it better is the first step toward caring for it wisely and using it well.

Acknowledgments

This book would not have been possible without the contributions of countless individuals whose work, support, and inspiration have shaped both the content and my own understanding of biological psychology.

First and foremost, I acknowledge the thousands of neuroscientists, psychologists, psychiatrists, and researchers whose painstaking work over more than a century has created our current understanding of the biological basis of behavior. From the early pioneers who first mapped brain functions to contemporary researchers using cutting-edge technologies, each has contributed to the magnificent edifice of knowledge that this book attempts to convey. While I cannot name them all, their collective effort deserves recognition and gratitude.

I am particularly indebted to the patients who have participated in research studies, often during their own times of suffering or vulnerability. Case studies of individuals with brain injuries, mental illness, or unusual conditions have provided invaluable insights into how the brain works. The remarkable people who have shared their experiences, allowed their brains to be studied, and contributed to scientific understanding deserve special recognition. Their willingness to advance knowledge that may help others exemplifies the best of human nature.

The educators who first introduced me to neuroscience and psychology sparked a fascination that has never diminished. Their ability to make complex concepts comprehensible and their enthusiasm for the subject matter inspired my own commitment to making biological psychology accessible to broader audiences. Good teaching is a gift that multiplies as students carry forward what they have learned and share it with others.

Colleagues and mentors who have challenged my thinking, pointed out gaps in my understanding, and encouraged deeper exploration have been invaluable. Scientific understanding advances through dialogue, debate, and the collision of ideas. Those willing to critique work constructively, to point out errors while acknowledging strengths, provide essential service to intellectual progress.

Librarians and information specialists who maintain the infrastructure allowing access to scientific literature deserve recognition. The ability to access thousands of research articles, to trace the development of ideas across decades, and to stay current with rapidly advancing fields depends on systems for organizing and disseminating scientific information. Those who maintain these systems enable the synthesis and integration that books like this attempt.

Friends and family who have supported this project through their patience, encouragement, and willingness to listen to enthusiastic explanations of neuroscience findings have provided essential support. Writing requires extended periods of focused work, and those who understood and supported this commitment made completion possible.

Readers of earlier drafts who provided feedback on clarity, accuracy, and accessibility helped improve the final product substantially. Identifying where explanations were unclear, where more examples were needed, and where technical language required translation has made the book more accessible to general audiences.

I acknowledge the limitations of any single work attempting to cover a field as vast as biological psychology. Choices about what to include and what to omit were necessary but inevitably imperfect. Topics that received brief mention could fill entire books themselves. Research findings are constantly emerging that may overturn current understanding. I have attempted to represent the current state of knowledge accurately while acknowledging uncertainties, but errors and oversimplifications surely remain. These are my responsibility alone.

Finally, I acknowledge the remarkable organ that made writing this book possible and that you are using to read it. The human brain, product of evolution and personal experience, represents the most complex structure we know. Understanding it better has been the work of a lifetime for many scientists and remains a work in progress. Contributing to broader understanding of the biological basis of human psychology has been a privilege and a joy.

To all who have contributed to this book directly or indirectly, through their research, teaching, support, or simply by asking interesting questions about how the brain creates mind thank you.

Introduction

WHY BIOLOGY MATTERS FOR UNDERSTANDING PSYCHOLOGY

Every moment of your conscious life every thought, emotion, sensation, and memory arises from biological processes in your brain. This fundamental truth connects the most intimate aspects of human experience with the physical world of neurons, neurotransmitters, and neural circuits. Understanding this connection is essential for understanding both normal psychological functioning and the disorders that cause suffering, for developing effective treatments for mental illness, for improving education and human development, and for making informed personal and societal decisions about issues ranging from criminal justice to cognitive enhancement.

The assertion that mental life has biological foundations may seem obvious, yet it represents a profound shift from views that dominated much of human history. For millennia, the mind was considered separate from the body, a spiritual entity temporarily housed in physical form. Thoughts and feelings were seen as fundamentally different from physical processes, operating according to different principles and perhaps not subject to scientific investigation at all.

The scientific revolution gradually eroded this dualism, showing that many phenomena once attributed to spiritual forces could be explained through natural processes. Yet the mind remained a holdout, seeming too complex, too personal, too uniquely human to reduce to biology. Even as medicine advanced and the brain's anatomy was mapped in detail, the connection between this physical organ and subjective mental experience remained mysterious.

The past century has witnessed a revolution in understanding this connection. Through convergence of evidence from multiple sources studies of brain-damaged patients, brain imaging of healthy individuals, pharmacological manipulation of neurotransmitters, genetic studies, and animal research neuroscience has established beyond doubt that mental processes are brain processes. There is no separate mind distinct from the brain; rather, mind emerges from brain as surely as digestion emerges from the stomach or circulation from the heart.

This biological understanding of psychology has profound implications. It explains why brain injuries produce specific psychological impairments damage to particular brain regions disrupts particular mental functions because those regions support those functions. It explains why drugs that alter brain chemistry affect mood and cognition neurotransmitters mediate mental processes, so changing neurotransmitter function changes mental states. It explains why mental illness can be treated with medications psychiatric disorders involve biological abnormalities that pharmacological interventions can address.

Understanding psychology's biological foundations also transforms how we view ourselves and others. Mental illness is recognized as genuine medical conditions involving measurable brain abnormalities rather than character flaws or personal failures. This reduces stigma and promotes seeking help. Individual differences in abilities, personality, and emotional reactivity are understood as reflecting both genetic variation and environmental influences on brain development rather than simply choices or willpower. This promotes more realistic expectations and better tailored education and interventions.

Yet biological psychology does not reduce human experience to mere mechanism or determinism. The brain is not a simple machine executing fixed programs but rather a remarkably plastic organ that reorganizes continuously in response to experience. Genes provide starting points that interact with environments in complex ways, with neither determining outcomes alone. Multiple levels of explanation from molecules to cultures are necessary for complete understanding, and higher levels involve emergent properties not reducible to lower levels even though they depend on them.

This book explores the biological foundations of psychology across multiple domains. We begin with foundations brain anatomy, neural communication, and nervous system organization establishing the structural and functional bases for mental life. We then examine specific psychological phenomena including emotion, stress, social bonding, memory, learning, and intelligence, exploring how biology creates these capacities while environmental factors shape their expression. We investigate sleep and its crucial functions, then examine mental health and illness, understanding conditions like depression as brain disorders amenable to biological interventions. Finally, we look toward the future, considering emerging technologies, unresolved questions, and ethical challenges.

Throughout, several themes recur. Biology and environment are inseparable, with genes requiring environments for expression and environments affecting biology through multiple mechanisms. The brain shows both stability and plasticity, with lasting characteristics coexisting with capacity for change. Multiple interacting brain systems support complex behaviors, with dysfunction in any component potentially causing problems. Individual differences are real and important, reflecting both biological variation and different experiences. Understanding mechanisms provides leverage for intervention, allowing more targeted and effective treatments than purely descriptive knowledge permits.

This biological approach to psychology does not diminish the richness of human experience but rather deepens appreciation for the sophisticated biological processes that create it. Love involves specific patterns of neural activity and hormone release, but this does not make it less meaningful or powerful. Memory involves changes in synaptic connections, but this does not make cherished memories less valuable. Depression involves altered brain chemistry and function, but this does not make the suffering less real or worthy of compassion and treatment.

Some worry that biological understanding threatens concepts like free will, moral responsibility, and human dignity. If all thoughts and actions arise from prior brain states determined by genes and experience, can we still be held responsible for choices? If abilities have biological limits, does this justify giving up on people who struggle? These concerns reflect misunderstandings of what biological psychology actually shows.

Biological psychology demonstrates that mental processes have physical bases, not that they are rigidly determined or that intervention is futile. The brain's plasticity means that experiences, including interventions, modify biology. Learning changes neural connections. Psychotherapy reorganizes neural circuits. Exercise promotes neurogenesis. The causation is bidirectional, with biology affecting psychology and psychology affecting biology. Understanding biological foundations empowers rather than constrains, revealing leverage points for positive change.

Regarding responsibility, biological explanations and moral accountability are compatible rather than contradictory. Explaining how brains make decisions does not eliminate the fact that persons make choices that have consequences for themselves and others. The legal and moral systems can incorporate biological understanding while maintaining accountability, perhaps becoming more sophisticated in distinguishing cases where biological factors substantially impaired capacity from cases where capacity was intact.

This book is written for general readers curious about the biological basis of psychology. No specialized background is assumed beyond basic education and interest in understanding how the brain creates mental life. Technical terms are explained when introduced, and concepts are illustrated with concrete examples and real cases. The goal is accessibility without oversimplification, accuracy without impenetrable jargon.

Different readers will approach this material with different backgrounds and interests. Students may use it as an introduction to biological psychology or as a complement to formal coursework. Educators might find insights for understanding how students learn and develop. Healthcare professionals could gain deeper understanding of the biological underpinnings of mental health conditions. Anyone curious about human nature, consciousness, or how brains work will find material for reflection.

The organization allows both sequential reading for comprehensive understanding and selective reading of specific topics. Early chapters provide foundations that later chapters build upon, but each chapter also strives to be comprehensible independently with cross-references to related material elsewhere. Readers can start at the beginning and progress through, or they can jump to topics of particular interest.

As you read, you will be using your brain to understand your brain a uniquely reflexive endeavor. The neurons processing these words, the memory systems encoding interesting information, the attention networks focusing on the page, and the emotional reactions you experience are all products of the biological processes described in these pages. This provides a special immediacy and relevance that distinguishes neuroscience from other sciences. We are simultaneously the observers and the observed, studying ourselves through the very capacities we seek to understand.

The human brain represents the most complex structure in the known universe. Comprehending even a fraction of how it works requires sustained effort. But the rewards are profound. Understanding the biological basis of psychology illuminates who we are, why we think and feel as we do, how we can improve our own functioning, and how we can help others who struggle. It provides foundations for evidence-based approaches to mental health, education, and human development. It inspires awe at the sophistication of the biological processes creating conscious experience.

Welcome to this exploration of the biology of the human mind. The journey will take you from the molecular mechanisms of neural communication to the systems-level organization supporting consciousness, from the genetic influences on personality to the environmental factors shaping development, from the neural basis of emotion to the biology of mental illness. Along the way, you will discover how three pounds of biological tissue creates everything you experience as mental life, and how understanding this process empowers you to protect brain health, maximize potential, and approach psychological challenges with knowledge and compassion.

Let us begin

1

THE MIND-BODY CONNECTION

The question of what makes us human has captivated philosophers, scientists, and thinkers for thousands of years. What separates our species from every other living creature on this planet? Is it our ability to reason, to feel complex emotions, to create art, to love, to dream? For most of human history, the mind was considered something separate from the body, a mysterious force that existed beyond the physical realm. Ancient Greeks believed the soul resided in the heart. Medieval scholars thought the brain was merely a cooling system for the blood. Religious traditions across cultures taught that consciousness was divine, untouchable by science, forever beyond human understanding.

But over the past century and a half, a revolution has been quietly unfolding in laboratories, hospitals, and research centers around the world. This revolution has fundamentally transformed our understanding of what it means to be human. Scientists have discovered that the mind is not separate from the body at all. Every thought you have ever had, every emotion you have ever felt, every memory you cherish, every dream that dances through your sleep, every moment of joy or sorrow or fear or love exists because of biology. Your mind is the product of approximately 86 billion neurons firing in intricate patterns inside your skull. Your personality is shaped by the chemical balance of neurotransmitters flooding across microscopic synapses. Your deepest feelings are orchestrated by hormones coursing through your bloodstream.

Your very sense of self emerges from the physical architecture of your brain.

This realization is both humbling and empowering. It is humbling because it reminds us that we are not supernatural beings floating above the natural world, but rather extraordinarily complex biological organisms subject to the same physical laws that govern all matter and energy in the universe. Yet it is empowering because understanding the biological basis of psychology gives us unprecedented tools to improve mental health, enhance learning, manage emotions, and unlock human potential in ways our ancestors could never have imagined.

Welcome to the field of biological psychology, also known as behavioral neuroscience, a discipline that bridges the gap between biology and psychology to answer one of the most fundamental questions we can ask: How does the three-pound mass of tissue inside your skull create the richness of human experience?

The Birth of a New Science

To understand where we are today, we must first understand how we got here. The journey toward biological psychology has been long, winding, and filled with both brilliant insights and terrible mistakes.

In ancient times, the relationship between brain and mind was poorly understood, if considered at all. The Greek physician Hippocrates, often called the father of medicine, was one of the first to propose that the brain was the seat of intelligence and emotion. Writing around 400 BCE, he boldly declared that from the brain, and from the brain only, arise our pleasures, joys, laughter, and jests, as well as our sorrows, pains, griefs, and tears. This was a radical idea at the time, contradicting popular belief that emotions originated in the heart or other organs.

However, Hippocrates' insight was not widely accepted. Aristotle, perhaps the most influential philosopher of the ancient world, rejected the idea entirely. He believed the brain was nothing more than a radiator, cooling the blood that was heated by the heart, which he considered the true center of sensation and thought. This view dominated Western thinking for nearly two thousand years, demonstrating how even the most brilliant minds can be mistaken when they lack the tools to observe nature accurately.

The Renaissance brought renewed interest in anatomy and the workings of the human body. Physicians began to dissect cadavers, mapping the structures of the brain with increasing detail. Yet even as anatomical knowledge grew, the relationship between brain structure and mental function remained mysterious. How could soft tissue produce something as intangible as a thought? How could chemical reactions create the sensation of love? How could electrical impulses generate consciousness itself?

The breakthrough began in the nineteenth century with a series of accidental discoveries and deliberate experiments that would forever change our understanding of the mind. One of the most famous cases involved a railroad construction foreman named Phineas Gage, whose story has become legendary in psychology and neuroscience textbooks around the world.

On September 13, 1848, Gage was working on the Rut land and Burlington Railroad in Vermont when a terrible accident occurred. An explosion drove a large iron rod completely through his skull, entering below his left cheekbone and exiting through the top of his head. Miraculously, Gage survived. He could walk and talk, and his memory remained intact. But something profound had changed. Before the accident, Gage had been responsible, hardworking, and well-liked. Afterward, he became impulsive, profane, and unable to follow through on plans. His friends said he was "no longer Gage."

This tragic case provided crucial evidence that specific regions of the brain control specific aspects of personality and behavior. The iron rod had destroyed parts of Gage's frontal lobes, the brain regions responsible for planning, impulse control, and social behavior. For the first time, scientists had clear proof that damage to particular brain areas could fundamentally alter who a person is. The mind, it seemed, was inseparable from the physical brain.

Around the same time, other researchers were making equally important discoveries. French physician Paul Broca studied patients who had lost the ability to speak following strokes. Through careful autopsies after these patients died, Broca discovered that they all had damage to the same region in the left frontal lobe, now known as Broca's area. This finding demonstrated that language production was localized to a specific part of the brain, further supporting the idea that mental functions have physical locations.

German physician Carl Wernicke soon identified another language area in the brain, this one involved in understanding speech rather than producing it. Patients with damage to Wernicke's area could speak fluently but their words made no sense, and they could not comprehend what others said to them. These discoveries established the principle of localization of function, the idea that different parts of the brain are responsible for different mental abilities.

As the twentieth century dawned, technology advanced rapidly, giving scientists new tools to peer inside the living brain. Electroencephalography allowed researchers to measure electrical activity in the brain through electrodes placed on the scalp. Later, brain imaging technologies like CT scans, MRI, PET scans, and functional MRI would allow scientists to watch the brain in action, observing which regions become active during different mental tasks. These tools transformed neuroscience from a field that relied primarily on studying brain-damaged patients to one that could investigate healthy brains engaged in thinking, feeling, remembering, and perceiving.

Meanwhile, biochemists were unraveling the chemical basis of neural communication. They discovered neurotransmitters, the chemical messengers that allow neurons to communicate with each other. They found that many mental illnesses were associated with imbalances in these chemicals. Depression was linked to low levels of serotonin and nor epinephrine. Schizophrenia was connected to excessive dopamine activity. Anxiety disorders involved disruptions in GABA, the brain's primary inhibitory neurotransmitter. These discoveries led to the development of psychiatric medications that could alleviate symptoms by adjusting brain chemistry, providing further evidence that mental states have biological foundations.

The revolution in genetics added yet another dimension to our understanding. Scientists discovered that many psychological traits and mental disorders run in families, suggesting genetic influences. The completion of the Human Genome Project in 2003 opened new frontiers in understanding how genes contribute to personality, intelligence, mental illness, and behavior. Researchers found that genes do not determine our psychology in any simple way, but rather interact with environmental factors in complex ways to shape who we become.

Today, biological psychology stands at an exciting crossroads. Advances in neuroscience, genetics, endocrinology, and related fields are providing unprecedented insights into how biology creates psychology. We can now observe individual neurons firing in real time, map the connections between billions of brain cells, identify genes associated with specific traits, and even modify brain activity using targeted interventions. The ancient question of how the mind relates to the body has been answered, though the full details of that relationship continue to unfold with each new discovery.

Why Biology Explains Behavior

Understanding that psychology is rooted in biology is not merely an academic exercise. It has profound practical implications for how we understand ourselves and treat mental health conditions. Consider the following scenarios, all of which demonstrate the inseparable connection between biology and psychology.

A college student struggling with depression visits a therapist, expecting to spend months analyzing childhood experiences and unconscious conflicts. Instead, after a thorough evaluation, she learns that her depressive symptoms may be related to a thyroid disorder. Her thyroid gland is not producing enough thyroid hormone, a condition called hypothyroidism. Because thyroid hormones play a crucial role in regulating energy, mood, and cognitive function, this biological problem was manifesting as psychological symptoms. Once her thyroid condition is treated with medication, her depression lifts. Her mind improves because her body improves.

A middle-aged man experiences sudden personality changes, becoming paranoid and suspicious of his wife and colleagues. His family assumes he is having a midlife crisis or experiencing stress at work. However, when his symptoms worsen and he begins having seizures, he undergoes a brain scan that reveals a tumor pressing on his temporal lobe. The tumor is affecting brain regions involved in emotion regulation and social perception, causing his behavioral changes. After surgery to remove the tumor, his personality returns to normal. His psychological symptoms were the direct result of a physical problem in his brain.

A young woman with severe anxiety has tried multiple therapies without success. Her racing thoughts, constant worry, and panic attacks make daily life almost unbearable. Finally, she tries a medication that increases GABA activity in her brain. GABA is the neurotransmitter that calms neural activity, counteracting excitatory signals. Within weeks of starting the medication, her anxiety diminishes significantly. The drug has not changed her life circumstances, her relationships, or her past experiences. It has simply altered her brain chemistry, demonstrating that her anxiety had a biological component that could be addressed biologically.

These examples illustrate a crucial principle: mental states are brain states. Every psychological phenomenon corresponds to some pattern of biological activity. This does not mean that psychology can be reduced entirely to biology, or that talking therapy, social support, and environmental changes are unimportant. Rather, it means that biological and psychological levels of explanation are complementary, not contradictory. Understanding the biological basis of behavior enhances rather than diminishes the richness of psychology.

The biological perspective also helps us move beyond outdated and harmful ways of thinking about mental illness. For centuries, people with psychological disorders were stigmatized, blamed for their conditions, or even punished. They were told to simply try harder, think more positively, or exercise more willpower. While lifestyle factors and cognitive strategies are indeed important, we now understand that many mental health conditions have strong biological components. Depression is not a character weakness or a failure of willpower. It involves measurable changes in brain structure and chemistry. Schizophrenia is not caused by bad parenting or moral failing. It is a neurodevelopmental disorder with genetic and neurochemical underpinnings. Anxiety disorders are not simply excessive worry that can be wished away. They involve hyperactive fear circuits in the brain that may require biological interventions to properly regulate.

Recognizing the biological basis of mental health conditions reduces stigma and promotes more effective treatment. If we understand that depression involves disrupted neurotransmitter systems, we can develop medications to correct those disruptions. If we know that anxiety involves an overactive amygdala, we can use therapies specifically designed to calm that brain region.

If we recognize that ADHD involves differences in dopamine regulation and frontal lobe function, we can create interventions targeted at those specific biological mechanisms.

At the same time, the biological perspective reminds us of the remarkable plasticity and adaptability of the brain. Unlike a computer with fixed hardware, the brain constantly rewires itself in response to experience. This phenomenon, called neuroplasticity, means that our biology is not our destiny. Yes, we are shaped by our genes, hormones, and brain structure. But we are also shaped by our experiences, choices, and environments, which in turn modify our biology. Learning a new skill creates new neural connections. Chronic stress can shrink the hippocampus, a brain region crucial for memory. Meditation can increase the thickness of the prefrontal cortex. Exercise promotes the growth of new neurons. The relationship between biology and psychology is bidirectional: biology influences psychology, and psychology influences biology.

The Scope and Promise of Biological Psychology

Biological psychology is an enormous field that encompasses many different levels of analysis and many different research methods. At the molecular level, researchers study how individual genes are expressed, how neurotransmitters bind to receptors, and how proteins within neurons change in response to experience. At the cellular level, scientists investigate how neurons generate electrical signals, how they connect with each other, and how networks of neurons process information. At the systems level, researchers examine how different brain regions work together to produce perception, memory, emotion, and behavior. At the behavioral level, scientists observe how biological factors influence what organisms do, from simple reflexes to complex social behaviors.

This multilevel approach is necessary because the mind emerges from biology at multiple scales simultaneously. Understanding a single neuron tells us something important, but it does not explain how millions of neurons working together produce consciousness. Understanding a single neurotransmitter helps us develop better medications, but it does not fully explain the complexity of depression or anxiety. Understanding a single gene associated with schizophrenia provides valuable insight, but it does not account for the many environmental and developmental factors that influence whether someone develops the disorder. Biological psychology integrates knowledge across all these levels to build a comprehensive understanding of how biology creates psychology.

The methods of biological psychology are equally diverse. Researchers use brain imaging to observe neural activity during mental tasks. They study patients with brain damage to understand what functions are lost when specific regions are injured. They examine the effects of drugs on behavior to infer how neurotransmitter systems work. They conduct experiments with animals to investigate brain mechanisms that cannot be studied in humans. They analyze genetic data to identify inherited influences on behavior. They measure hormones to understand their effects on mood and cognition. They record electrical activity from individual neurons to understand how information is coded in the brain. Each method has strengths and limitations, and converging evidence from multiple methods provides the strongest conclusions.

The promise of biological psychology extends far beyond the laboratory. Understanding the biological basis of behavior has transformative potential for medicine, education, criminal justice, and society as a whole. In medicine, biological psychology is leading to better treatments for mental illness, addiction, neurodegenerative diseases, and brain injuries. In education, insights about how the brain learns are informing teaching methods that align with how students actually acquire and retain information. In criminal justice, knowledge about brain development is raising important questions about culpability and rehabilitation, especially for adolescents whose frontal lobes are still maturing. In society at large, biological psychology is fostering greater compassion and understanding for people whose behavior differs from the norm, whether due to mental illness, neurodevelopmental differences, or brain injuries.

Yet with this promise comes responsibility. As we gain the ability to measure, predict, and potentially modify the biological basis of behavior, we face profound ethical questions. Should we use genetic testing to identify people at risk for mental illness? Should we give children medications to enhance their academic performance? Should we use brain scans in courtrooms to determine criminal responsibility? Should we allow people to use neurotechnology to enhance their cognitive abilities beyond normal levels? These questions do not have easy answers, and biological psychology must be pursued with careful attention to ethical implications and potential for misuse.

The Journey Ahead

This book will take you on a journey through the biological foundations of human psychology. We will explore the structure and function of the brain, examining how different regions contribute to different aspects of mental life. We will investigate the chemical and electrical signals that allow neurons to communicate, and how these signals create thoughts and feelings. We will examine how hormones influence mood, motivation, and behavior. We will look at the role of genes in shaping personality, intelligence, and mental health. We will explore how evolution has shaped the human mind, leaving us with ancient instincts that sometimes clash with modern life. We will investigate the biological basis of emotions, from the fear that protects us from danger to the love that binds us to others. We will examine how the brain learns and remembers, and what goes wrong in conditions like Alzheimer's disease. We will look at mental health through a biological lens, understanding depression, anxiety, addiction, and other conditions as disorders with physical substrates that can be measured and treated.

Throughout this journey, we will encounter fascinating case studies, groundbreaking experiments, and cutting-edge research that reveals the intimate connection between brain and mind. You will learn how a stroke patient who cannot recognize faces can still sense whether someone is familiar. You will discover why teenagers take more risks than adults, and why this is not simply poor judgment but rather reflects the ongoing development of their frontal lobes. You will understand why jetlag makes you feel terrible, and how your body's internal clock regulates sleep and wakefulness. You will learn why stress can make you sick, and how chronic activation of the stress response damages both brain and body. You will explore why some people are naturally more anxious or more extroverted than others, and how both genes and environment contribute to these differences.

By the end of this book, you will have a deep appreciation for the biological foundations of psychology. You will understand that you are not a ghost in a machine, a soul trapped in a body, or a mind separate from matter. You are an integrated whole, a biological organism whose psychology emerges from the complex interactions of genes, neurons, hormones, and environment. This understanding does not diminish the wonder of human experience. If anything, it enhances it. The fact that evolution and biology have produced beings capable of love, creativity, morality, and self-awareness is nothing short of miraculous.

The ancient question of how mind relates to body has been answered: they are one. But that answer opens up countless new questions about the details of that relationship. How exactly do neurons create consciousness? How do genes and environment interact to shape who we become? How can we harness our understanding of biology to reduce suffering and enhance well-being? These questions drive biological psychology forward, and the answers promise to transform our understanding of what it means to be human.

The journey begins with the brain itself, that wrinkled mass of tissue that contains you, everything you have ever known, everyone you have ever loved, and every dream you have ever had. In the next chapter, we will explore the architecture of this remarkable organ, examining how its structure gives rise to the rich tapestry of human mental life.

2

THE ARCHITECTURE OF THE HUMAN BRAIN

If you could hold a human brain in your hands, you might be surprised by how unimpressive it looks at first glance. It weighs about three pounds, roughly the same as a small cantaloupe. Its surface is wrinkled and folded like a walnut, with a texture resembling firm tofu or custard. It has a grayish-pink color, though preserved brains often appear more gray or tan. There are no flashing lights, no visible movements, no obvious signs of the extraordinary activity occurring within. Yet this modest-looking organ is the most complex structure we know of in the entire universe.

The human brain contains approximately 86 billion neurons, each of which can form thousands of connections with other neurons, creating a network of staggering complexity. If you tried to count all the neural connections in a single human brain at a rate of one per second, it would take you more than three million years to finish. These billions of neurons communicate through trillions of synapses, generating patterns of electrical and chemical activity that somehow produce everything you experience as consciousness. Your brain is reading these words right now, extracting meaning from abstract symbols, relating new information to what you already know, perhaps generating emotional reactions or memories as you read, all while simultaneously regulating your heartbeat, breathing, digestion, and countless other bodily functions without any conscious effort on your part.

Understanding the architecture of the brain is essential to understanding how biology creates psychology. Just as you cannot fully appreciate how a city functions without knowing its layout, neighborhoods, and infrastructure, you cannot understand the mind without knowing the brain's structure. The brain is not a uniform mass but rather a collection of specialized regions, each with distinct functions, all working together in intricate coordination. Some regions are ancient in evolutionary terms, shared with reptiles and even fish. Others are more recent developments, expanded dramatically in primates and reaching their greatest elaboration in humans. This layered architecture reflects the brain's evolutionary history and helps explain both our remarkable cognitive abilities and our more primitive emotional responses.

The Major Divisions: A Brain in Three Parts

Neuroscientists often divide the brain into three major parts based on their evolutionary development and general functions. The oldest part, sometimes called the reptilian brain or hind brain, controls basic survival functions. The middle part, or mid brain, coordinates movement and relays sensory information. The newest part, the fore brain, handles complex thinking, planning, and consciousness. This division is somewhat simplified, as these regions extensively interact and share functions, but it provides a useful framework for understanding brain organization.

The hind brain sits at the base of the skull where the brain connects to the spinal cord. It includes the medulla oblongata, the pons, and the cerebellum. The medulla oblongata is perhaps the most critical structure for immediate survival. It controls vital functions that must continue without interruption: breathing, heart rate, and blood pressure. Damage to the medulla is usually fatal because these functions cease. The medulla operates entirely without conscious control. You cannot decide to stop your heart or cease breathing indefinitely. These functions are too important to be left to conscious decision-making, which can be distracted or impaired. Instead, the medulla monitors oxygen levels, carbon dioxide concentrations, and other critical variables, adjusting your breathing and heart rate automatically to maintain homeostasis.

Just above the medulla lies the pons, a bulge on the brain stem that serves as a bridge connecting different parts of the brain. The pons plays important roles in sleep, arousal, and facial expressions. It also helps relay information between the cerebellum and the cerebral cortex. The name "pons" literally means bridge in Latin, reflecting its role in connecting brain regions that need to communicate with each other.

The cerebellum extends from the back of the brain stem, looking somewhat like a smaller brain attached to the larger one. For centuries, scientists underestimated the cerebellum's importance, viewing it primarily as a structure for balance and coordination. While the cerebellum does indeed play a crucial role in motor control, recent research has revealed that it does much more. The cerebellum contains more neurons than the rest of the brain combined, despite occupying only about ten percent of the brain's volume. This remarkable density of neurons suggests the cerebellum performs extensive computational processing.

The cerebellum acts like a quality control center for movement, comparing intended movements with actual movements and making rapid corrections when they do not match. When you reach for a coffee cup, your cerebral cortex initiates the movement based on where you see the cup and where you want your hand to go. But slight errors inevitably occur in executing this plan. Perhaps your arm extends a bit too far or your fingers open too wide. The cerebellum detects these discrepancies in milliseconds and sends corrective signals to adjust your movement in real time, ensuring your hand arrives at exactly the right position with exactly the right grip to grasp the cup smoothly. This happens so quickly and automatically that you never consciously notice the corrections being made.

Damage to the cerebellum produces characteristic symptoms that reveal its functions. People with cerebellar damage often exhibit ataxia, a lack of coordination that makes movements jerky and imprecise. They may stumble when walking, overshoot when reaching for objects, and struggle with tasks requiring fine motor control like buttoning a shirt or writing. Their movements retain their basic structure but lose their smoothness and accuracy. Interestingly, cerebellar damage also affects non-motor functions. Some patients develop difficulty with timing and rhythm, struggle with certain cognitive tasks, and experience changes in emotional regulation. These findings suggest the cerebellum may perform similar error-correction and fine-tuning functions for cognitive and emotional processes, though researchers are still working to understand these non-motor roles fully.

Above the hind brain lies the mid brain, a relatively small region that serves crucial functions in movement, motivation, and sensory processing. The mid brain contains structures that control eye movements, allowing you to track moving objects and shift your gaze to new locations. It includes the substantia nigra, a cluster of neurons that produce dopamine and play essential roles in movement initiation and reward processing. Degeneration of the substantia nigra causes Parkinson's disease, a movement disorder characterized by tremors, rigidity, and difficulty initiating movement. The mid brain also contains the ventral tegmental area, another dopamine-producing region that is central to motivation, pleasure, and addiction. When you experience something rewarding, whether eating delicious food, receiving a compliment, or using an addictive drug, neurons in the ventral tegmental area become active, releasing dopamine that creates feelings of pleasure and motivates you to repeat the behavior.

The fore brain is the largest and most complex division, containing structures responsible for our most sophisticated mental abilities. The fore brain includes the thalamus, hypothalamus, limbic system, and cerebral cortex. Each of these components deserves detailed attention because they play central roles in creating human psychology.

The Thalamus: The Brain's Grand Central Station

Deep within the center of the brain lies the thalamus, a pair of egg-shaped structures that serve as the brain's central relay station. Nearly all sensory information from your body passes through the thalamus before reaching the cerebral cortex where conscious perception occurs. Visual information from your eyes, sounds from your ears, touch sensations from your skin, and taste information from your tongue all flow through the thalamus, which processes, organizes, and routes this information to appropriate cortical regions.

The thalamus is not merely a passive relay station but actively shapes the information passing through it. It filters sensory data, suppressing some signals while amplifying others, helping to determine what reaches your conscious awareness. When you focus intently on reading this book, the thalamus helps reduce your awareness of background noises, the feeling of clothes against your skin, and other sensory information that is not immediately relevant. This filtering is essential because the world contains far more sensory information than your conscious mind could possibly process. Without the thalamus selectively routing information, you would be overwhelmed by the constant barrage of sensations from your environment and body.

The thalamus also plays important roles in regulating consciousness, sleep, and alertness. Damage to the thalamus can produce profound effects on awareness, sometimes leaving people in persistent vegetative states or minimally conscious states where they cycle between sleep-like and wake-like periods but show little meaningful responsiveness to their environment. The precise mechanisms by which the thalamus contributes to consciousness remain a topic of intense research, but its role is clearly essential.

One striking exception to the rule that sensory information passes through the thalamus is the sense of smell. Olfactory information travels directly from the nose to the olfactory cortex and then to emotional centers in the brain, bypassing the thalamus entirely. This direct route may explain why smells can trigger powerful emotional memories so readily. The scent of fresh-baked cookies might instantly transport you back to your grandmother's kitchen, or the smell of a particular perfume might evoke memories of a past relationship. These olfactory memories often carry strong emotional content because smell information reaches emotional brain centers without first being processed by higher cognitive areas.

The Hypothalamus: Small Structure, Enormous Influence

Just below the thalamus sits the hypothalamus, a structure no larger than an almond yet absolutely critical for survival. Despite its small size, the hypothalamus regulates an astonishing array of basic drives and bodily functions. It controls hunger and thirst, body temperature, sexual behavior, sleep-wake cycles, stress responses, and the release of hormones from the pituitary gland. The hypothalamus essentially links the nervous system to the endocrine system, translating neural signals into hormonal messages that regulate physiology throughout the body.

The hypothalamus monitors your body's internal state constantly, detecting deviations from optimal levels and initiating corrective responses. When your body temperature rises, the hypothalamus triggers sweating and dilates blood vessels in your skin to release heat. When temperature drops, it causes shivering and constricts skin blood vessels to conserve warmth. When blood sugar falls, the hypothalamus generates hunger, motivating you to eat. When you become dehydrated, it creates thirst and triggers the release of hormones that tell your kidneys to conserve water. These regulatory processes, collectively called homeostasis, keep your body's internal environment stable despite constant changes in external conditions.

The hypothalamus also orchestrates the body's response to stress through its control of the hypothalami-pituitary-adrenal axis, commonly abbreviated as the HPA axis. When you encounter something threatening or stressful, neurons in the hypothalamus release a hormone called corticotropin-releasing hormone. This hormone travels to the nearby pituitary gland, stimulating it to release adrenocorticotropic hormone into the bloodstream. This second hormone then reaches the adrenal glands, which sit atop your kidneys, causing them to release cortisol, the primary stress hormone. Cortisol mobilizes energy stores, suppresses non-essential functions like digestion and reproduction, and prepares your body for action. This cascade of hormonal signals allows the hypothalamus to coordinate a whole-body response to stress, demonstrating how a tiny brain structure can have enormous physiological effects.

Different regions within the hypothalamus regulate different motivated behaviors. The lateral hypothalamus promotes eating; damage to this area produces loss of appetite and weight loss. The ventromedial hypothalamus signals satiety; damage here causes overeating and obesity. The suprachiasmatic nucleus serves as the brain's master clock, regulating circadian rhythms by responding to light information from the eyes and coordinating daily cycles of sleepiness and alertness. The preoptic area regulates body temperature and aspects of sexual behavior. These specialized regions allow the hypothalamus to fine-tune multiple biological drives simultaneously, balancing competing needs and maintaining overall physiological stability.

The Limbic System: The Emotional Brain

Surrounding the thalamus and hypothalamus lies a collection of interconnected structures collectively known as the limbic system. This term, derived from the Latin word for border or edge, refers to structures that form a border around the brain stem. The limbic system plays central roles in emotion, motivation, and memory. Its major components include the amygdala, hippocampus, and cingulate cortex, though the exact boundaries of the limbic system remain somewhat debated among neuroscientists.

The amygdala is an almond-shaped structure tucked deep within the temporal lobe, one on each side of the brain. Despite its small size, the amygdala wields tremendous influence over emotional experience and behavior, particularly fear and anxiety. The amygdala acts as the brain's threat detection system, constantly scanning your environment for potential dangers. When it identifies something threatening, it triggers rapid physiological responses before conscious awareness catches up. This is why you might jump at a sudden noise before you consciously recognize what made the sound, or why you might feel your heart racing when watching a scary movie even though you know rationally that you are safe.

Research has revealed fascinating details about how the amygdala processes threats. Information about potential dangers reaches the amygdala through two pathways, sometimes called the low road and the high road. The low road carries crude sensory information directly from the thalamus to the amygdala in a fraction of a second, allowing ultra-fast responses to threats. This pathway sacrifices accuracy for speed; it may trigger fear in response to a curved stick that vaguely resembles a snake. The high road routes sensory information through the cerebral cortex first, where it undergoes detailed analysis before reaching the amygdala. This pathway is slower but more accurate, eventually allowing you to realize that the stick is not actually a snake. This dual-pathway system makes evolutionary sense. False alarms are annoying but relatively harmless, while failing to respond quickly to genuine threats can be fatal. Better to jump at a hundred sticks than to fail to avoid one snake.

The amygdala does more than just detect threats. It also assigns emotional significance to experiences, helping to create emotional memories that guide future behavior. When something important happens, whether positive or negative, the amygdala enhances memory formation for that event. This is why emotional experiences tend to be remembered more vividly than neutral ones. You likely remember where you were during emotionally significant events like your first kiss, receiving acceptance to college, or learning about a major news event, but you probably cannot recall what you had for lunch three Tuesdays ago. The amygdala strengthens memory for emotional events because they are more likely to be important for survival and future decision-making.

Damage to the amygdala produces striking changes in emotional processing. People with bilateral amygdala damage often show reduced fear responses and difficulty recognizing fear in other people's facial expressions. One well-studied patient known as SM has damage to both amygdalae due to a rare genetic condition. She shows virtually no fear in situations that would terrify most people. Researchers have taken her to haunted houses, let her handle snakes and spiders, and shown her frightening films, yet she reports little to no fear. Interestingly, SM can still experience other emotions like happiness and sadness, and she can even feel fear in response to panic-inducing drugs that affect brain chemistry directly, suggesting that the amygdala is specifically necessary for fear triggered by external threats rather than for the basic capacity to experience fear.

Adjacent to the amygdala lies the hippocampus, a seahorse-shaped structure that plays an indispensable role in forming new long-term memories. Without a functioning hippocampus, you could not remember what happened five minutes ago, where you parked your car this morning, or what you learned in yesterday's lecture. The hippocampus converts temporary short-term memories into more permanent long-term memories through a process called consolidation, and it creates the spatial and temporal context that allows you to remember when and where events occurred.

The critical role of the hippocampus in memory became dramatically clear through the famous case of patient Henry Molaison, known for decades as simply HM to protect his privacy. In 1953, at age 27, HM underwent experimental brain surgery to treat severe epilepsy. Surgeons removed large portions of his medial temporal lobes, including most of his hippocampus, on both sides of his brain. The surgery successfully reduced his seizures, but it produced devastating and unexpected consequences for his memory.

After the surgery, HM could no longer form new long-term memories. He could hold information in mind for a few minutes through active rehearsal, but as soon as his attention shifted, the information vanished completely. He could read the same magazine repeatedly without recognizing it. He could meet the same person dozens of times but would have no memory of previous encounters. He could not remember what year it was or recognize that he had aged. His memories from before the surgery remained largely intact, but his life from that point forward became frozen in an eternal present, each moment vanishing as soon as it passed. HM's case provided definitive proof that the hippocampus is essential for forming new long-term memories, and his generous participation in research over more than five decades made him one of the most important patients in the history of neuroscience.

The hippocampus does not store memories permanently. Rather, it acts like a librarian who catalogs and organizes information before filing it in long-term storage elsewhere in the brain, particularly in the cerebral cortex. When you try to remember something, the hippocampus helps retrieve scattered pieces of information from various cortical areas and binds them together into a coherent memory. Over time, as memories are repeatedly recalled and re consolidated, they may become less dependent on the hippocampus and more firmly established in cortical networks. This gradual transfer from hippocampus to cortex explains why hippocampal damage typically affects recent memories more severely than remote memories. HM could remember events from his childhood and young adulthood reasonably well, but he could not remember anything that happened after his surgery.

Beyond its role in declarative memory, the type of memory for facts and events that you can consciously recall, the hippocampus is also crucial for spatial navigation and creating cognitive maps of your environment. Research with both animals and humans has shown that the hippocampus contains place cells, neurons that fire when an organism is in a particular location in space. These neurons essentially create an internal GPS system, allowing you to know where you are and how to navigate to where you want to go. London taxi drivers, who must memorize thousands of streets and routes, show enlarged hippocampi compared to other people, demonstrating the brain's remarkable ability to expand structures in response to intensive use.

The cingulate cortex wraps around the corpus callosum, the massive bundle of fibers connecting the brain's two hemispheres. The cingulate cortex has multiple subdivisions with different functions. The anterior cingulate cortex, located toward the front, plays important roles in decision-making, error detection, and regulating emotions. It becomes active when you must choose between conflicting options or when you make mistakes. It also appears to be involved in the emotional aspects of pain; people with damage to the anterior cingulate cortex may still feel the physical sensation of pain but report that it no longer bothers them emotionally. The posterior cingulate cortex, located further back, contributes to memory retrieval and self-referential thinking, the type of thought involved when you think about yourself and your own mental states.

Together, the structures of the limbic system work in concert to create emotional experiences, form memories, and guide motivated behavior. They represent evolutionarily old parts of the brain that we share with many other mammals, which explains why animals like dogs and rats clearly experience emotions and can learn from experience even though they lack the sophisticated reasoning abilities that depend on more recently evolved brain regions.

The Cerebral Cortex: The Crowning Achievement

The largest and most conspicuous part of the human brain is the cerebral cortex, the wrinkled outer layer that gives the brain its characteristic appearance. The cortex is a sheet of neural tissue about two to three millimeters thick, roughly the thickness of a few stacked coins. If you could unfold and flatten the cortex, it would cover an area of about 2,500 square centimeters, roughly the size of a large napkin. To fit this much surface area inside the skull, the cortex folds extensively, creating the distinctive wrinkles and grooves you see on a brain's surface. The raised folds are called gyri, while the grooves between them are called sulci. Deeper grooves are called fissures.

The cerebral cortex is divided into two hemispheres, left and right, connected by the corpus callosum. Each hemisphere is further divided into four lobes named for the skull bones that cover them: the frontal lobe, parietal lobe, temporal lobe, and occipital lobe. Each lobe handles different aspects of perception, thought, and action, though they constantly communicate and cooperate to produce unified mental experience.

The occipital lobe, located at the back of the brain, is devoted primarily to vision. Visual information from the eyes travels through the thalamus to the primary visual cortex, located at the very back of the occipital lobe. This region creates basic visual representations based on patterns of light and dark, edges, and simple shapes. More complex visual processing occurs in surrounding visual association areas that recognize objects, faces, motion, and color. Damage to the occipital cortex can cause blindness even when the eyes themselves function normally, demonstrating that seeing depends as much on the brain as on the eyes. Some patients with occipital damage experience blindsight, a peculiar condition where they claim to be unable to see but can nonetheless respond to visual stimuli at above-chance levels when forced to guess, suggesting that some visual information processing occurs outside conscious awareness.

The temporal lobe extends along the sides of the brain, roughly behind your temples. The temporal lobe processes auditory information, with the primary auditory cortex receiving sound information from the ears via the thalamus. Surrounding areas analyze complex sounds including speech and music. The temporal lobe also houses critical regions for memory, including the hippocampus and parts of the cortex involved in storing knowledge about objects, facts, and concepts. Damage to the left temporal lobe often impairs language comprehension, as in Wernicke's aphasia, while damage to the right temporal lobe may impair recognition of melodies and environmental sounds. The temporal lobes also contribute to face recognition, with some regions showing particularly strong responses to faces compared to other visual stimuli.

The parietal lobe sits near the top and back of the brain, behind the frontal lobe and above the temporal and occipital lobes. The parietal lobe integrates sensory information from different modalities, creating coherent representations of your body and its relationship to space. The primary somatosensory cortex, located at the front of the parietal lobe just behind a major groove called the central sulcus, receives touch, temperature, and pain information from the body. This region is organized as a map of the body surface, with adjacent body parts represented in adjacent cortical areas, though the map is distorted rather than proportional. Areas with high tactile sensitivity like the hands, lips, and tongue receive disproportionately large representations, while less sensitive areas like the back and legs receive relatively small representations.

Beyond the primary somatosensory cortex, the parietal lobe contains association areas that perform more complex spatial processing. These regions help you grasp objects accurately, navigate through environments, and maintain awareness of where your body parts are positioned. Damage to parietal areas can produce bizarre symptoms that reveal these functions. Some patients develop neglect syndrome, where they become unaware of one side of space, typically the left side following right parietal damage. These patients may eat food from only the right side of their plate, draw only the right half of a clock, or fail to dress the left side of their body, not because they cannot move or sense the left side but because it simply does not exist in their conscious awareness. Other patients develop apraxia, an inability to perform purposeful movements despite having normal strength and sensation, suggesting that parietal regions store representations of how to execute complex action sequences.

The frontal lobe, which occupies the entire front portion of the brain, is the last to fully mature during development and shows the greatest expansion in humans compared to other species. The frontal lobe handles motor control, planning, decision-making, impulse control, and many aspects of personality and social behavior. The rear portion of the frontal lobe contains the primary motor cortex, which sends commands to muscles throughout the body. Like the somatosensory cortex, the motor cortex is organized as a body map, with body parts requiring fine motor control like the hands and face receiving disproportionately large representations.

In front of the motor cortex lies the premotor cortex, which plans movements before they are executed, and the supplementary motor area, which coordinates sequences of movements and appears particularly important for self-initiated actions as opposed to responses to external stimuli. Even further forward lies the prefrontal cortex, the brain region most associated with distinctively human mental abilities. The prefrontal cortex occupies a larger proportion of total brain volume in humans than in any other species, and it continues developing well into the twenties, explaining in part why adolescents often struggle with impulse control, planning, and considering long-term consequences of their actions.

The prefrontal cortex performs executive functions, high-level cognitive processes that control and coordinate other mental abilities. These include working memory, the ability to hold information in mind and manipulate it; cognitive flexibility, the ability to switch between different tasks or mental sets; and inhibitory control, the ability to suppress inappropriate or impulsive responses. The prefrontal cortex allows you to pursue long-term goals despite short-term temptations, consider multiple perspectives before making decisions, and regulate your emotions rather than being controlled by them.

Different regions within the prefrontal cortex contribute to different executive functions. The dorsolateral prefrontal cortex plays important roles in working memory, planning, and cognitive control. The ventromedial prefrontal cortex integrates emotional information with decision-making, helping you make choices that consider both logical analysis and emotional significance. The orbitofrontal cortex processes reward value and punishment information, allowing you to learn from feedback and adjust behavior accordingly. Damage to these regions produces characteristic impairments. People with dorsolateral prefrontal damage may struggle to organize their thoughts, plan ahead, or resist distractions. Those with ventromedial or orbitofrontal damage may make poor decisions despite intact reasoning abilities, suggesting that good decision-making requires integrating emotion with logic rather than relying on pure rationality.

The case of Phineas Gage, mentioned in Chapter 1, provides a vivid illustration of frontal lobe function. The iron rod that pierced Gage's skull damaged his left frontal lobe, particularly orbitofrontal and ventromedial regions. After the accident, Gage retained his intelligence, memory, and ability to speak, but his personality changed dramatically. He became impulsive, socially inappropriate, and unable to pursue long-term goals. He could understand what he should do but could not consistently follow through. His case demonstrates that the frontal lobes are essential not just for thinking but for translating thoughts into appropriate, goal-directed behavior.

Left Brain, Right Brain: Separating Myth from Reality

The fact that the brain has two hemispheres has captured public imagination and generated considerable mythology. Popular culture often describes people as left-brained or right-brained, supposedly reflecting whether they are more logical and analytical or more creative and intuitive. While the two hemispheres do have some specialized functions, the popular version vastly oversimplifies and often misrepresents the actual science.

The two hemispheres are indeed structurally asymmetric in subtle ways, and they do show some functional specialization, a phenomenon called lateralization. The most striking lateralization involves language. For about 95 percent of right-handed people and about 70 percent of left-handed people, language functions are strongly lateralized to the left hemisphere. Broca's area and Wernicke's area, the two major language regions discussed earlier, are located in the left hemisphere for most people. Damage to the left hemisphere typically produces more severe language impairments than equivalent damage to the right hemisphere.

However, the right hemisphere is not uninvolved in language. It processes prosody, the emotional melody of speech, and it understands metaphors and jokes that require going beyond literal word meanings. The right hemisphere also excels at spatial processing, facial recognition, and processing the global configuration of visual scenes, while the left hemisphere shows advantages for detail-oriented processing and sequential information. These differences are relative rather than absolute; both hemispheres contribute to most mental functions.

In normal brains, the two hemispheres communicate extensively through the corpus callosum, sharing information rapidly so that you experience a unified mental life rather than having two separate minds inside one skull. The popular notion of people being left-brained or right-brained, with one hemisphere dominant for all functions, has no scientific support. Brain imaging studies consistently show that both hemispheres activate during virtually all mental tasks. You do not use one hemisphere for math and the other for art. You do not have a logical left brain battling with a creative right brain. You have a unified brain with two hemispheres that specialize somewhat differently but work together seamlessly.

The most revealing evidence about hemispheric specialization comes from split-brain patients, individuals who have had their corpus callosum surgically severed to treat severe epilepsy. With the main connection between hemispheres cut, each hemisphere processes information somewhat independently. Researchers can present information selectively to one hemisphere by briefly showing it in one visual field, since each hemisphere primarily receives information from the opposite side of space. These studies have revealed that the two hemispheres can have different knowledge, perceptions, and even preferences, though the significance and interpretation of these findings remain debated.

In daily life, split-brain patients function remarkably normally, showing few obvious impairments. This demonstrates the brain's remarkable ability to compensate for injuries and the fact that subcortical connections between the hemispheres, though less extensive than the corpus callosum, still allow some coordination. The subtle deficits that do appear in careful testing provide fascinating insights into how the brain integrates information, but they do not suggest that normal people have dramatically different mental capacities in each hemisphere or that education should target specific hemispheres.

The Living Architecture

This tour through brain anatomy reveals an organ of breathtaking complexity and elegant organization. From the primitive survival functions of the brainstem to the sophisticated reasoning of the prefrontal cortex, the brain's architecture reflects both our evolutionary heritage and our unique human capacities. Each structure contributes something essential to mental life, yet no structure works in isolation. Understanding the brain requires appreciating both specialization and integration, both localization and distributed processing.

The brain's physical structure constrains what minds can do, but it does not rigidly determine our psychology. The same basic architecture can produce enormous individual variation in personality, ability, and behavior, shaped by genes, development, experience, and culture. Moreover, the brain's architecture is not static. Throughout life, experience modifies neural connections, strengthening some pathways and pruning others. The physical brain you have today differs in countless ways from the brain you had last year or will have next year.

3

NEURONS – THE LANGUAGE OF THE BRAIN

Every thought you have ever had, every sensation you have ever experienced, every movement you have ever made, and every emotion you have ever felt depends on the activity of neurons. These remarkable cells are the fundamental units of the nervous system, the biological equivalent of transistors in a computer or bricks in a building. Yet neurons are far more dynamic and complex than any component in human technology. They communicate with each other through intricate electrical and chemical signals, forming networks that process information, store memories, generate consciousness, and control behavior. Understanding how individual neurons work and how they communicate is essential for understanding how biology creates psychology.

The human brain contains approximately 86 billion neurons, a number so large it defies easy comprehension. If you tried to count every neuron in your brain at a rate of one per second, working twenty-four hours a day without breaks, it would take you nearly three thousand years to finish. Yet the number of neurons, impressive as it is, tells only part of the story. Each neuron can connect with thousands of other neurons, creating a network of staggering complexity. Conservative estimates suggest the human brain contains at least

100 trillion synapses, the connection points where neurons communicate with each other. This means your brain has more connections than there are stars in the Milky Way galaxy, more connections than there are trees on Earth, and more connections than there are words in all the books ever written.

This chapter explores how neurons generate and transmit information, how they communicate with each other through chemical messengers called neurotransmitters, and how these microscopic processes give rise to everything we experience as mental life. We will examine the structure of neurons, the electrical signals they produce, the chemical communication that occurs between them, and how drugs, diseases, and experiences alter these fundamental processes. By understanding neurons, we gain insight into how the brain stores memories, why antidepressant medications take weeks to work, how addictive drugs hijack normal brain function, and why brain injuries produce specific patterns of impairment.

The Anatomy of a Neuron

Neurons come in many shapes and sizes, but most share a common basic structure consisting of three main parts: the cell body, dendrites, and an axon. Each component plays a specific role in receiving, processing, and transmitting information.

The cell body, also called the soma, contains the nucleus and other cellular machinery needed to keep the neuron alive and functioning. The nucleus houses the genetic material that provides instructions for making proteins, the molecular workhorses that carry out most cellular functions. Surrounding the nucleus are mitochondria, often called the powerhouses of the cell, which generate the energy needed to fuel the neuron's activities. Neurons are extraordinarily energy-hungry cells. Although the brain comprises only about two percent of body weight, it consumes roughly twenty percent of the body's oxygen and glucose. This enormous energy demand reflects the constant electrical and chemical activity required to maintain neural function.

Extending from the cell body are dendrites, branching structures that look somewhat like the branches of a tree. The word dendrite comes from the Greek word for tree, reflecting this branching appearance. Dendrites serve as the neuron's receiving stations, collecting incoming signals from other neurons. A single neuron may have thousands of dendrites, each forming connections with other neurons and receiving chemical messages from them. The dendrites integrate all these incoming signals, combining them in complex ways that determine whether the neuron will generate its own signal to pass along to other neurons.

The structure of dendrites is not fixed but changes throughout life in response to experience. Learning new skills, forming new memories, and adapting to new environments all involve changes in dendritic structure. Dendrites can grow new branches, extending their reach to form connections with additional neurons. They can also retract branches, pruning away connections that are no longer needed. This structural plasticity allows the brain to rewire itself based on experience, forming the biological foundation for learning and memory. Studies have shown that animals raised in enriched environments with toys, social companions, and opportunities for exploration develop more elaborate dendritic trees than animals raised in impoverished environments, demonstrating how experience literally shapes brain structure at the cellular level.

Extending from the cell body in the opposite direction from dendrites is the axon, a single long fiber that carries signals away from the cell body toward other neurons. While dendrites are typically short and numerous, each neuron has only one axon, which can vary tremendously in length. Some axons are less than a millimeter long, connecting nearby neurons within the same brain region. Others stretch for remarkable distances. The axons that control your toes extend from neurons in your spinal cord all the way down your legs, a distance that can exceed three feet in tall individuals. These long-distance axons allow different brain and body regions to communicate with each other, coordinating activity across the entire nervous system.

Many axons are wrapped in myelin, a fatty insulating substance that dramatically increases the speed of signal transmission. Myelin consists of specialized cells called glial cells that wrap around axons in layers, creating an insulating sheath much like the plastic coating on electrical wires. This insulation prevents electrical signals from leaking out and allows them to travel much faster than they could in unmyelinated axons. Myelin does not cover axons continuously but rather forms segments separated by small gaps called nodes of Ranvier. The electrical signal jumps from one node to the next, a process called saltatory conduction that is much faster than continuous propagation along the axon. Myelinated axons can transmit signals at speeds up to 120 meters per second, fast enough to travel the length of a football field in less than a second, while unmyelinated axons typically transmit signals at only about one meter per second.

The importance of myelin becomes tragically clear in demyelinating diseases like multiple sclerosis, where the immune system mistakenly attacks and destroys myelin. As myelin deteriorates, signal transmission slows and becomes unreliable, producing symptoms that depend on which axons are affected. Patients may experience vision problems, muscle weakness, coordination difficulties, cognitive impairment, and fatigue. The variability and unpredictability of symptoms in multiple sclerosis reflect the fact that demyelination can affect different brain and spinal cord regions in different patients and at different times. Understanding the role of myelin has led to treatments aimed at reducing immune attacks on myelin and promoting myelin repair, offering hope for patients with these devastating disorders.

At the end of the axon are terminal branches that end in small swellings called terminal buttons or synaptic terminals. These terminals contain tiny sacs called synaptic vesicles, which store neurotransmitters, the chemical messengers that neurons use to communicate with each other. When an electrical signal reaches the terminal button, it triggers the release of neurotransmitters into the small gap between neurons, allowing the signal to cross from one neuron to the next. This gap, called a synapse, is where the real magic of neural communication occurs.

The Electrical Language of Neurons

Neurons communicate through two types of signals: electrical signals that travel within a neuron and chemical signals that pass between neurons. Understanding how neurons generate and transmit electrical signals is essential for understanding how information flows through the nervous system.

Every neuron maintains an electrical charge difference across its cell membrane, with the inside of the neuron being negatively charged relative to the outside. This charge difference, called the resting potential, typically measures about negative seventy millivolts. The resting potential exists because of an unequal distribution of electrically charged ions, particularly sodium, potassium, chloride, and various negatively charged proteins, across the cell membrane. The neuron actively maintains this charge difference through protein pumps embedded in the membrane that continuously move ions against their concentration gradients, expending considerable energy in the process.

When a neuron receives signals from other neurons through its dendrites, these signals cause small changes in the electrical charge across the membrane. Some signals, called excitatory signals, make the inside of the neuron slightly less negative, pushing it toward the threshold for generating an action potential. Other signals, called inhibitory signals, make the inside of the neuron more negative, pushing it away from threshold. The neuron integrates all these incoming signals, adding up excitatory and inhibitory inputs to determine whether the combined effect reaches threshold.

If the combined input is strong enough to reach threshold, typically around negative fifty-five millivolts, the neuron generates an action potential, a rapid electrical signal that travels down the axon. The action potential represents an all-or-nothing event, much like firing a gun. Just as a gun either fires or does not fire depending on whether you pull the trigger hard enough, a neuron either generates an action potential or does not, depending on whether the input reaches threshold. Once initiated, the action potential always has the same magnitude regardless of how far above threshold the triggering stimulus was. A barely threshold stimulus produces the same action potential as a much stronger stimulus.

The action potential occurs through a carefully orchestrated sequence of events involving protein channels in the cell membrane that allow specific ions to flow across the membrane. When the membrane reaches threshold, voltage-gated sodium channels open, allowing sodium ions to rush into the neuron, driven by both their concentration gradient and the electrical gradient. This influx of positive sodium ions causes the inside of the neuron to become rapidly more positive, swinging from about negative fifty-five millivolts to about positive forty millivolts in just one millisecond. This rapid change in electrical charge is the rising phase of the action potential.

Almost immediately after the sodium channels open, they begin to close, and voltage-gated potassium channels open. Potassium ions flow out of the neuron, driven by their concentration gradient. This outflow of positive charges returns the neuron toward its negative resting potential during the falling phase of the action potential. In fact, the potassium channels stay open slightly too long, causing the membrane potential to overshoot and become even more negative than the resting potential before returning to baseline. This brief hyperpolarization makes the neuron temporarily less excitable, creating a refractory period during which it is difficult or impossible to generate another action potential.

The action potential is a self-regenerating signal that propagates down the axon without diminishing in strength. When the membrane at one location depolarizes during an action potential, it causes adjacent regions of the membrane to depolarize, triggering action potentials in those regions. This process continues along the entire length of the axon, much like a falling row of dominoes where each domino knocks over the next. In myelinated axons, the action potential jumps from one node of Ranvier to the next, dramatically speeding up transmission while conserving energy.

Neurons encode information intensity not through the size of action potentials, which are always the same magnitude, but through the frequency of action potentials. A weak stimulus might trigger a few action potentials per second, while a strong stimulus might trigger hundreds per second. When you gently touch a surface, sensory neurons in your skin fire at low frequencies. When you press harder, the same neurons fire at higher frequencies, and your brain interprets this higher firing rate as greater pressure. This frequency coding allows neurons to transmit information about stimulus intensity using identical action potentials.

The discovery of how neurons generate and transmit electrical signals represents one of the great achievements of neuroscience. Early scientists had hints that electricity was involved in neural function. In the eighteenth century, Luigi Galvani discovered that electrical stimulation could cause dead frog legs to twitch, suggesting that electricity somehow controlled muscle movement. In the nineteenth century, researchers developed techniques to measure electrical activity in nerves, confirming that neural communication involved electrical signals. But the detailed mechanisms remained mysterious until the mid-twentieth century, when British scientists Alan Hodgkin and Andrew Huxley used a giant axon from a squid to work out the ionic mechanisms underlying the action potential. Their work, which earned them the Nobel Prize in 1963, provided the foundation for all modern understanding of neural electrical activity.

Chemical Communication Between Neurons

When an action potential reaches the terminal button at the end of an axon, it initiates a remarkable process of chemical communication. The terminal button sits very close to the dendrite or cell body of another neuron but does not actually touch it. A small gap called the synaptic cleft, measuring only about twenty to forty nanometers wide, separates the two neurons. For perspective, this gap is roughly one two-thousandth the width of a human hair. Yet this tiny space is where neurons pass information from one to another through chemical messengers called neurotransmitters.

The arrival of an action potential at the terminal button causes voltage-gated calcium channels to open, allowing calcium ions to flow into the terminal. The influx of calcium triggers synaptic vesicles, tiny sacs containing neurotransmitter molecules, to fuse with the cell membrane and release their contents into the synaptic cleft. This process, called exocytosis, dumps thousands of neurotransmitter molecules into the gap between neurons. The neurotransmitters diffuse rapidly across the synaptic cleft, taking only about one millisecond to cross the gap and reach the receiving neuron, called the postsynaptic neuron.

On the surface of the postsynaptic neuron are specialized proteins called receptors that recognize and bind specific neurotransmitters, much like a lock recognizes a specific key. When neurotransmitter molecules bind to receptors, they cause changes in the postsynaptic neuron, typically by opening or closing ion channels in the membrane. Some receptors, called ionotropic receptors, are themselves ion channels that open when neurotransmitter binds. Others, called metabotropic receptors, activate more complex chemical cascades inside the postsynaptic neuron that can have diverse and long-lasting effects.

Different neurotransmitters produce different effects depending on which receptors they activate. Some neurotransmitters are primarily excitatory, making the postsynaptic neuron more likely to generate an action potential. Others are primarily inhibitory, making the postsynaptic neuron less likely to fire. Still others have more complex modulatory effects, adjusting the responsiveness of neurons to other inputs. The same neurotransmitter can even have different effects in different brain regions or on different receptor types, adding another layer of complexity to neural communication.

After neurotransmitters bind to receptors and trigger changes in the post-synaptic neuron, they must be cleared from the synaptic cleft to prevent continuous stimulation. This clearance occurs through several mechanisms. Some neurotransmitters are broken down by enzymes in the synaptic cleft. Others are taken back up into the presynaptic neuron through transporter proteins, a process called reuptake, where they can be repackaged into vesicles for future use. Some neurotransmitters drift away from the synapse and are taken up by glial cells. The speed and efficiency of neurotransmitter clearance is crucial for proper neural function, and many drugs work by interfering with these clearance mechanisms.

Scientists have identified more than one hundred different substances that function as neurotransmitters, though a smaller number account for most neural communication. Each neurotransmitter has specific functions, acts on specific receptors, and is found in specific brain pathways. Understanding these neurotransmitters is essential for understanding both normal brain function and the effects of drugs that alter brain chemistry.

Major Neurotransmitters and Their Functions

Glutamate is the primary excitatory neurotransmitter in the brain, involved in learning, memory, and general brain activation. Nearly every neuron in the brain responds to glutamate, and excessive glutamate activity can be toxic to neurons. During strokes, when blood flow to brain regions is interrupted, damaged neurons release massive amounts of glutamate, which overexcites neighboring neurons and contributes to the spread of damage. This excitotoxicity has led researchers to investigate drugs that might block excessive glutamate activity to limit stroke damage, though finding safe and effective treatments has proven challenging.

GABA, short for gamma-aminobutyric acid, is the primary inhibitory neurotransmitter, acting as a brake on neural activity throughout the brain. GABA counterbalances glutamate, preventing excessive excitation and maintaining stable brain function. Anxiety disorders often involve insufficient GABA activity, leaving the brain in a state of excessive excitation and arousal. Many anti-anxiety medications, including benzodiazepines like Valium and Xanax, work by enhancing GABA activity, effectively increasing inhibition and calming overactive neural circuits. Alcohol also enhances GABA activity, explaining its calming and sedating effects and also explaining why alcohol withdrawal can cause dangerous hyper excitability, including seizures.

Dopamine plays crucial roles in movement, motivation, reward, and pleasure. Dopamine-producing neurons are concentrated in several midbrain areas, particularly the substantia nigra and ventral tegmental area, which send projections throughout the brain. The substantia nigra projects primarily to motor control regions, and degeneration of these neurons causes Parkinson's disease, a movement disorder characterized by tremors, rigidity, and difficulty initiating movement. The ventral tegmental area projects to regions involved in motivation and reward, creating what is often called the brain's reward circuit. When you experience something pleasurable, whether eating delicious food, winning money, receiving social approval, or using addictive drugs, dopamine neurons in the ventral tegmental area become active, creating positive feelings and motivating you to repeat the rewarding behavior.

The powerful role of dopamine in reward and motivation makes it central to understanding addiction. Virtually all addictive drugs, from cocaine to heroin to methamphetamine, increase dopamine activity in reward circuits, though through different mechanisms. Cocaine blocks dopamine reuptake, allowing dopamine to accumulate in synapses. Amphetamines cause dopamine release and block reuptake. Heroin indirectly increases dopamine by removing inhibitory brakes on dopamine neurons. The resulting surge of dopamine creates intense pleasure and powerfully reinforces drug-taking behavior. With repeated drug use, the brain adapts by reducing dopamine receptors or changing how dopamine neurons respond, leading to tolerance and requiring larger doses to achieve the same effect. Meanwhile, normal pleasures become less satisfying because the adapted brain responds less strongly to natural rewards, creating a state where the addicted person feels pleasure primarily from the drug and little pleasure from anything else.

Serotonin influences mood, appetite, sleep, and many other functions. Serotonin neurons originate in the raphe nuclei, clusters of neurons in the brainstem, and project throughout the brain and spinal cord. Low serotonin activity has been linked to depression, though the relationship is more complex than once thought. Many antidepressant medications, including selective serotonin reuptake inhibitors like Prozac, Zoloft, and Paxil, work by blocking serotonin reuptake, allowing serotonin to remain in synapses longer and accumulate to higher levels. These medications take several weeks to produce their full therapeutic effects, suggesting they work not just by immediately increasing serotonin but by triggering slower adaptive changes in how neurons respond to serotonin.

Serotonin also regulates appetite and satiety. Drugs that increase serotonin activity often reduce appetite, which is why some serotonin-boosting medications cause weight loss as a side effect. Conversely, drugs that block certain serotonin receptors can increase appetite and cause weight gain. The serotonin system also plays important roles in regulating aggression, impulsivity, and sleep. People with low serotonin activity sometimes show increased aggression and impulsive behavior, and some violent criminals have been found to have reduced serotonin function.

Norepinephrine, also called noradrenaline, is involved in arousal, alertness, and stress responses. Norepinephrine neurons are concentrated in a small brainstem region called the locus coeruleus, which sends widespread projections throughout the brain. When you are startled or encounter something important or threatening, locus coeruleus neurons fire intensely, releasing norepinephrine that increases alertness and focuses attention. The norepinephrine system is part of the brain's alarm system, preparing you to respond to challenges or dangers.

Many antidepressant medications affect norepinephrine as well as serotonin. Some medications, like venlafaxine and duloxetine, block the reuptake of both serotonin and norepinephrine. Others, like bupropion, primarily affect norepinephrine and dopamine rather than serotonin. The existence of multiple types of antidepressants that work through different neurotransmitter systems suggests that depression is not simply a serotonin deficiency but rather involves complex imbalances across multiple neurochemical systems.

Acetylcholine was the first neurotransmitter discovered and plays diverse roles throughout the nervous system. At the junction between motor neurons and muscles, acetylcholine triggers muscle contraction. In the brain, acetylcholine is crucial for attention, learning, and memory. Alzheimer's disease, the most common cause of dementia, involves degeneration of acetylcholine-producing neurons, particularly in a structure called the nucleus basalis. As these neurons die, acetylcholine levels decline, contributing to the progressive memory loss and cognitive decline characteristic of the disease. Some medications for Alzheimer's disease work by blocking the enzyme that breaks down acetylcholine, allowing it to accumulate to higher levels and partially compensating for the loss of acetylcholine-producing neurons. These drugs do not cure Alzheimer's disease or stop its progression, but they can temporarily improve symptoms in some patients.

Acetylcholine is also the target of many toxins and drugs. Botulinum toxin, produced by bacteria and known commercially as Botox, blocks acetylcholine release at the neuromuscular junction, causing paralysis. In cosmetic applications, small amounts injected into facial muscles prevent contraction, reducing wrinkles. In larger amounts, botulinum toxin can cause respiratory paralysis and death, making it one of the most poisonous substances known. Conversely, some insecticides and nerve gases work by blocking the enzyme that breaks down acetylcholine, causing excessive accumulation that leads to uncontrollable muscle contractions, seizures, and death.

Endorphins are the brain's natural opiates, neurotransmitters that reduce pain and create feelings of pleasure and well-being. The word endorphin combines "endogenous," meaning produced within the body, and "morphine," the powerful painkiller derived from opium poppies. Endorphins bind to the same receptors in the brain that morphine and heroin activate, though endorphins are released naturally in response to pain, stress, exercise, and other stimuli. The runner's high, a sense of euphoria sometimes experienced during prolonged exercise, appears to result partly from endorphin release, though other neurochemical systems contribute as well.

The discovery of endorphins in the 1970s solved a long-standing mystery: why does the brain have receptors for morphine, a plant chemical? The answer is that these receptors did not evolve to respond to morphine but rather to the brain's own opioid-like chemicals. Morphine and heroin happen to have molecular structures similar enough to endorphins to activate the same receptors, but far more powerfully than endorphins do. This explains both the intense euphoria produced by opioid drugs and their high addiction potential. The brain's reward system responds to opioids as if they were natural rewards magnified many times over, creating powerful motivation to continue using the drug.

How Drugs Affect Neurotransmission

Understanding how neurotransmitters work explains how psychoactive drugs, medications that affect mental processes, produce their effects. Every drug that alters mood, perception, or consciousness does so by changing neurotransmitter function in some way. Some drugs mimic neurotransmitters, binding to their receptors and activating them. Others block receptors, preventing neurotransmitters from binding. Still others affect neurotransmitter synthesis, storage, release, or clearance. The specificity of these effects explains why different drugs have different psychological effects and why the same drug can have multiple effects by acting on multiple neurotransmitter systems.

Agonist drugs facilitate neurotransmitter function, either by mimicking the neurotransmitter and binding to its receptors or by increasing neurotransmitter availability. Nicotine is an agonist at acetylcholine receptors, binding to a specific subtype called nicotinic receptors and activating them. This activation in brain reward circuits produces the mild pleasure associated with smoking and contributes to tobacco addiction. Morphine and heroin are agonists at opioid receptors, mimicking endorphins but producing much stronger activation. L-DOPA, used to treat Parkinson's disease, is converted into dopamine in the brain, compensating for the loss of dopamine-producing neurons.

Antagonist drugs block neurotransmitter function by occupying receptors without activating them, preventing the natural neurotransmitter from binding. Many antipsychotic medications used to treat schizophrenia are dopamine antagonists, blocking dopamine receptors and reducing excessive dopamine activity thought to contribute to hallucinations and delusions.

Naloxone is an opioid antagonist used to reverse opioid overdoses. It rapidly displaces morphine or heroin from opioid receptors, blocking their effects and restoring normal breathing in overdose victims. However, because naloxone blocks the receptors so completely, it also precipitates immediate and severe withdrawal in people physically dependent on opioids, making the treatment lifesaving but extremely unpleasant.

Some drugs work by blocking neurotransmitter reuptake, as mentioned earlier with antidepressants. By preventing neurotransmitters from being taken back into the presynaptic neuron, these drugs allow neurotransmitters to remain in the synapse longer, accumulating to higher concentrations and producing stronger or more prolonged effects. Cocaine blocks the reuptake of dopamine, norepinephrine, and serotonin, though its effects on dopamine appear most important for its rewarding properties and addiction potential. The intense but short-lived euphoria produced by cocaine results from a rapid surge in dopamine levels, followed by a crash as dopamine becomes depleted and the drug wears off.

Other drugs affect neurotransmitter synthesis or storage. Reserpine, an old medication for high blood pressure, depletes neurotransmitters by interfering with their storage in synaptic vesicles. This depletion can cause severe depression, a side effect that provided early evidence linking neurotransmitters to mood and helped establish the neurochemical basis of depression. Conversely, L-DOPA increases dopamine synthesis by providing the raw material neurons use to manufacture dopamine, partially compensating for the loss of dopamine-producing neurons in Parkinson's disease.

The complexity of neurotransmitter systems means that drugs rarely have simple, single effects. Most psychoactive drugs affect multiple neurotransmitter systems, producing a constellation of effects that can be therapeutic, undesirable, or both. Antidepressants that increase serotonin can cause sexual dysfunction as a side effect because serotonin also regulates sexual function. Antipsychotic medications that block dopamine to reduce hallucinations can cause movement problems resembling Parkinson's disease because dopamine is also essential for normal movement control. Understanding these multiple effects helps explain why finding effective medications with tolerable side effects remains challenging and why the same medication may work well for some patients but not others.

Neural Plasticity: The Changing Brain

One of the most remarkable discoveries in neuroscience is that neural connections are not fixed but constantly change throughout life in response to experience. This capacity for change, called neural plasticity or neuroplasticity, allows the brain to learn, adapt, and recover from injury. Every time you form a new memory, learn a new skill, or adapt to a new environment, your brain physically changes at the level of individual neurons and synapses.

Learning involves changes in synaptic strength, the effectiveness with which one neuron influences another. When two neurons are repeatedly active at the same time, the connection between them tends to strengthen, a process summarized by the phrase "neurons that fire together wire together." This activity-dependent strengthening of synapses, called long-term potentiation, is thought to be a fundamental mechanism underlying learning and memory. Conversely, synapses that are rarely used tend to weaken or disappear, a process called synaptic pruning that refines neural circuits by eliminating unnecessary connections.

Structural changes accompany these functional changes. Learning complex skills causes dendrites to grow new branches, expanding their capacity to form connections with other neurons. Brain imaging studies show that London taxi drivers, who must memorize thousands of streets and routes, have enlarged hippocampi compared to other people, with the enlargement correlating with years of navigation experience. Musicians show structural differences in brain regions involved in motor control and auditory processing, with changes corresponding to the age at which they began training. These findings demonstrate that experience literally reshapes brain structure, not just during early development but throughout life.

The brain shows greatest plasticity during critical periods early in development when neural circuits are being established. Young children learn languages, develop perceptual abilities, and acquire skills with ease that becomes more difficult in adulthood. This enhanced plasticity allows rapid learning but also creates vulnerability to deprivation or abnormal experiences that can permanently alter brain development. Children raised in severely deprived environments, lacking normal social interaction and stimulation, often show persistent cognitive and emotional difficulties that reflect altered brain development during critical periods.

However, adult brains retain substantial plasticity, particularly when motivated by necessity or intensive practice. People who become blind often develop enhanced auditory and tactile abilities as brain regions normally devoted to vision become recruited for processing other sensory modalities. Stroke patients can recover lost functions through rehabilitation that takes advantage of surviving brain regions assuming new functions. Musicians and athletes demonstrate that intensive practice can produce remarkable changes in brain structure and function even in adulthood, expanding representations of body parts and refining neural circuits involved in skilled performance.

Understanding neural plasticity has profound implications for education, rehabilitation, and mental health treatment. If experience shapes brain structure and function, then carefully designed experiences might help people overcome learning disabilities, recover from brain injuries, or modify maladaptive patterns of thinking and behavior. This insight underlies many modern therapeutic approaches, from cognitive behavioral therapy that aims to rewire thought patterns to constraint-induced movement therapy that forces stroke patients to use impaired limbs to promote brain reorganization.

The brain's capacity for change offers hope that biology is not destiny and that intervention can reshape neural circuits in beneficial ways.

Beyond Neurons: The Supporting Cast

While neurons receive most attention in discussions of brain function, they are not the only cells in the nervous system. Glial cells, once thought to simply provide structural support for neurons, actually outnumber neurons and perform diverse and essential functions. The name glia comes from the Greek word for glue, reflecting the early belief that these cells merely held neurons in place. Modern research reveals that glia actively participate in brain function in ways that are only beginning to be understood.

Astrocytes, star-shaped glial cells, regulate the chemical environment around neurons, controlling concentrations of ions and neurotransmitters in the extracellular space. They remove excess neurotransmitters and potassium ions that could interfere with neural signaling. Astrocytes also provide neurons with nutrients, forming connections between blood vessels and neurons that transfer glucose and other essential molecules. Recent research suggests astrocytes may actively participate in information processing, responding to neural activity and influencing synaptic transmission, though their precise computational roles remain debated.

Oligodendrocytes in the central nervous system and Schwann cells in the peripheral nervous system produce the myelin that wraps around axons and speeds signal transmission. Each oligodendrocyte can myelinate segments of multiple axons, while each Schwann cell myelinates only a single axon segment. Myelin formation continues well into adolescence and even adulthood, particularly in frontal brain regions involved in complex reasoning and impulse control. This protracted myelination may contribute to the cognitive and behavioral changes that occur during adolescence and early adulthood as neural communication becomes faster and more efficient.

Microglia serve as the brain's immune cells, constantly surveying the neural environment for signs of damage, infection, or debris that needs to be cleared away. When microglia detect problems, they become activated, changing shape and releasing chemical signals that trigger inflammatory responses. This immune function protects the brain from pathogens and helps remove damaged cells, but excessive or chronic microglial activation may contribute to neurodegenerative diseases and other brain disorders. Recent research suggests microglia also play important roles in normal development by pruning excess synapses, refining neural circuits through a process that involves selectively eliminating connections that are weak or unnecessary.

The recognition that glia actively contribute to brain function rather than merely supporting neurons has transformed neuroscience. Researchers now investigate how glial dysfunction contributes to neurological and psychiatric disorders, how glia participate in information processing and memory formation, and how glia-neuron interactions shape brain development and plasticity. Understanding the brain requires understanding not just neurons but the entire cellular ecosystem in which they operate.

From Molecules to Mind

This exploration of neurons, neurotransmitters, and neural communication reveals how physical and chemical processes create mental life. Every sensation, thought, emotion, and action corresponds to patterns of neural activity, to neurotransmitters binding to receptors, to changes in synaptic strength, to electrical signals propagating through neural networks. The molecular mechanisms described in this chapter are not merely biological details but the very substance of psychology, the physical processes that generate consciousness, store memories, create emotions, and produce behavior.

Understanding neurons provides a foundation for understanding how drugs work, why brain injuries produce specific impairments, how learning changes the brain, and how mental disorders arise from disrupted neural function. It explains why antidepressants take weeks to work, not because they slowly increase neurotransmitter levels, which happens within hours, but because they gradually trigger adaptive changes in how neurons respond to those neurotransmitters. It explains why addiction is so difficult to overcome, as drugs create powerful changes in dopamine circuits that persist long after drug use stops. It explains why memories are not stored in single neurons but in patterns of synaptic connections distributed across many neurons, making memory both robust and vulnerable to disruption.

Yet for all we have learned about neurons, profound mysteries remain. How does the activity of billions of individual neurons give rise to unified conscious experience? How are memories encoded in patterns of synaptic strengths and later retrieved? How do neural networks perform the computational operations underlying perception, decision-making, and thought? These questions drive continuing neuroscience research, gradually revealing how microscopic processes in individual cells combine to create the full richness of mental life.

4

THE NERVOUS SYSTEM – COMMAND AND CONTROL

The brain does not exist in isolation. Despite its extraordinary complexity and central importance, the brain is only one component of a larger system that extends throughout the entire body. This system, called the nervous system, consists of billions of neurons organized into networks that gather information from the environment, process that information, make decisions, and execute actions. Understanding the complete nervous system reveals how the brain connects to the world, how it controls the body, and how mental processes depend on constant communication between brain and body.

The nervous system divides into two major parts based on location and function. The central nervous system, abbreviated CNS, consists of the brain and spinal cord. These structures are called central because they occupy the midline of the body and serve as the main processing centers for information. The peripheral nervous system, abbreviated PNS, includes all neural tissue outside the brain and spinal cord. The peripheral nervous system carries information from sensory receptors to the central nervous system and carries commands from the central nervous system to muscles and glands. These two divisions work together seamlessly, creating an integrated system that allows you to perceive, think, feel, and act.

This chapter explores how the nervous system is organized, how different components work together, and how this organization shapes psychological experience. We will examine the spinal cord and its role in reflexes and communication between brain and body. We will investigate the peripheral nervous system and how it gathers sensory information and controls movement. We will explore the autonomic nervous system, which regulates internal organs and bodily states largely without conscious control. Finally, we will examine how the nervous system responds to challenges and stress, preparing the body for action when danger threatens and restoring calm when threats pass.

The Spinal Cord: Information Highway and Reflex Center

The spinal cord is a cylinder of neural tissue about eighteen inches long and roughly as thick as your index finger that extends from the base of the brain down through the vertebral column, the series of bones commonly called the spine or backbone. The vertebral column provides essential protection for the spinal cord, encasing it in bone much as the skull protects the brain. Despite this protection, spinal cord injuries occur when trauma fractures vertebrae and damages the delicate neural tissue within, often with devastating consequences.

The spinal cord serves two critical functions. First, it acts as the primary communication pathway between the brain and the rest of the body. Sensory information from the skin, muscles, joints, and internal organs travels up the spinal cord to the brain, while motor commands from the brain travel down the spinal cord to control muscles throughout the body. Second, the spinal cord can process certain types of information locally and generate responses without involving the brain, producing what are called spinal reflexes.

The spinal cord is not simply a bundle of nerve fibers but contains substantial neural processing machinery. Gray matter in the center of the spinal cord contains neuron cell bodies and synapses where information processing occurs. White matter surrounding the gray matter consists primarily of myelinated axons organized into tracts that carry information up and down the cord. Ascending tracts carry sensory information toward the brain, while descending tracts carry motor commands from the brain toward muscles. Different tracts carry different types of information, with some devoted to touch, others to pain and temperature, and still others to various aspects of motor control.

The organization of these tracts has important clinical implications. When the spinal cord is injured, the specific symptoms depend on which tracts are damaged and at what level the damage occurs. Injury to ascending sensory tracts eliminates sensation from body regions below the injury site. Damage to descending motor tracts causes paralysis of muscles controlled by spinal cord segments below the injury. Because tracts carrying different types of information occupy different locations within the spinal cord, some injuries selectively affect certain modalities while sparing others. A person might lose pain and temperature sensation on one side of the body while retaining touch sensation, or might lose voluntary movement while retaining reflexes, depending on exactly which parts of the spinal cord are damaged.

Spinal reflexes demonstrate that complex behaviors can occur without brain involvement. The most familiar example is the stretch reflex, which causes a muscle to contract when it is suddenly stretched. When a physician taps your knee with a reflex hammer, the impact briefly stretches the quadriceps muscle in your thigh. Sensory neurons embedded in the muscle detect this stretch and send signals to the spinal cord. These sensory neurons connect directly to motor neurons that control the quadriceps, forming a simple two-neuron circuit. When the sensory neuron is activated by stretch, it excites the motor neuron, which triggers muscle contraction, causing your leg to kick forward. This entire sequence, from stretch to kick, takes only about fifty milliseconds and occurs entirely within the spinal cord without any brain involvement.

The stretch reflex serves important functions in maintaining posture and coordinating movement. When you stand upright, gravity constantly pulls your body downward, stretching postural muscles. The stretch reflex automatically causes these muscles to contract, resisting gravity and keeping you upright without requiring conscious effort or attention. During walking, the stretch reflex helps coordinate leg movements by automatically adjusting muscle contractions based on how much muscles are being stretched. These automatic adjustments allow smooth, coordinated movement without burdening the brain with controlling every muscle contraction.

Another important spinal reflex is the withdrawal reflex, which pulls your hand or foot away from painful stimuli. When you accidentally touch something hot or sharp, sensory neurons carrying pain information activate spinal circuits that simultaneously contract muscles that pull the limb away from danger and relax muscles that would oppose this withdrawal. This coordinated pattern of muscle activation and relaxation occurs automatically through spinal circuits before pain information reaches your brain and before you consciously decide to withdraw. Only after your hand has already jerked away does pain sensation reach your conscious awareness, demonstrating that the nervous system can initiate protective actions faster than conscious processing would allow.

While spinal reflexes occur automatically without brain involvement, the brain can modulate them. Descending pathways from the brain can strengthen or weaken reflex responses depending on behavioral context. An athlete anticipating a starting gun might suppress the startle reflex that would normally cause flinching, while someone walking on a slippery surface might enhance reflexes that help maintain balance. This modulation allows reflexes to be adjusted appropriately for different situations while still occurring rapidly and automatically when needed.

The importance of the spinal cord becomes tragically clear when it is damaged. Spinal cord injury is classified by level and completeness. Level refers to where along the spinal cord damage occurs, typically described by which vertebra is affected. Injuries higher in the spinal cord affect more of the body because they interrupt communication between the brain and all spinal segments below the injury. Complete injuries sever all neural pathways at a particular level, causing total loss of sensation and voluntary movement below the injury. Incomplete injuries damage some but not all pathways, producing variable patterns of preserved and lost function.

An injury at the neck level, called a cervical injury, can cause quadriplegia, paralysis of all four limbs and the trunk. Injuries at progressively lower levels cause paraplegia, paralysis of the legs and lower trunk, with higher injuries affecting more of the trunk and lower injuries affecting primarily the legs. Beyond paralysis and sensory loss, spinal cord injuries disrupt autonomic regulation of body functions, causing problems with blood pressure control, temperature regulation, bladder and bowel function, and sexual function. These widespread effects reflect the spinal cord's role not just in movement and sensation but in coordinating physiological regulation throughout the body.

Current treatments for spinal cord injury focus on preventing additional damage, promoting whatever recovery is possible, and helping people adapt to disabilities through rehabilitation and assistive technology. Researchers are investigating potential therapies to promote neural regeneration and restore function, including stem cell treatments, drugs that promote axon growth, electrical stimulation, and robotic devices that might bypass damaged neural pathways. While complete restoration of function remains elusive, these approaches offer hope for improving outcomes after spinal cord injury.

The Peripheral Nervous System: Connecting Brain to Body

The peripheral nervous system encompasses all neural tissue outside the brain and spinal cord, including sensory neurons that detect information about the world and the body, and motor neurons that control muscles. The peripheral nervous system divides into somatic and autonomic divisions based on what they control and whether their actions are typically conscious or unconscious.

The somatic nervous system includes sensory neurons that carry information from sensory receptors to the central nervous system, and motor neurons that carry commands from the central nervous system to skeletal muscles. Skeletal muscles are the muscles attached to bones that move your limbs, torso, and head in response to voluntary commands. When you decide to reach for a cup, turn your head, or take a step, motor neurons in the somatic nervous system carry commands from your brain and spinal cord to the appropriate skeletal muscles.

Sensory neurons in the somatic nervous system have specialized receptors that respond to different types of stimulation. Mechanoreceptors respond to mechanical distortion such as touch, pressure, vibration, and stretching. Different mechanoreceptors have different properties that make them suited for detecting different aspects of touch. Some respond rapidly to changes in pressure, making them ideal for detecting vibration and texture. Others respond continuously during sustained pressure, allowing you to feel continuous contact with objects. Still others detect stretching of skin or muscles, providing information about body position and movement.

Thermoreceptors respond to temperature, with separate receptors for warmth and cold. These receptors do not simply measure absolute temperature but respond to temperature changes and differences between skin temperature and object temperature. This is why a room-temperature object can feel cold if your hand is warm or feel warm if your hand is cold. It also explains why jumping into a cold pool feels shocking at first but becomes more tolerable as your skin temperature adjusts, reducing the temperature difference that activates cold receptors.

Nociceptors respond to potentially harmful stimuli that could damage tissue, generating sensations we experience as pain. Different nociceptors respond to different types of potentially harmful stimulation, including intense mechanical pressure, extreme temperatures, and chemicals released by damaged tissue. Pain signals travel to the spinal cord through two types of nerve fibers with different properties. Fast fibers covered in myelin transmit sharp, localized pain that tells you precisely where injury occurred and prompts immediate withdrawal. Slow fibers without myelin transmit dull, aching, poorly localized pain that persists after initial injury. This dual pain system explains the common experience of hurting yourself and first feeling a sharp pain that quickly gives way to a deeper, more prolonged ache.

Proprioceptors located in muscles, tendons, and joints detect body position and movement, providing what is called proprioceptive or kinesthetic information. This sense allows you to know where your limbs are located even with your eyes closed, to coordinate complex movements, and to maintain balance. Proprioception operates largely outside conscious awareness, automatically guiding movement and posture. You become aware of proprioception primarily when it is disrupted, such as when your foot falls asleep and you temporarily lose the normal sense of where it is positioned, making walking feel awkward and uncoordinated.

Motor neurons in the somatic nervous system extend from the spinal cord to skeletal muscles, with their cell bodies located in the ventral, or front, portion of the spinal cord gray matter. When these motor neurons fire action potentials, they release acetylcholine at the neuromuscular junction, the synapse between motor neuron and muscle fiber. The acetylcholine binds to receptors on the muscle fiber, triggering a cascade of events that causes muscle contraction. Each motor neuron connects to multiple muscle fibers, with the motor neuron and all the muscle fibers it controls called a motor unit.

The number of muscle fibers in a motor unit varies depending on the precision required. Muscles that perform fine, delicate movements like those controlling eye movements or finger movements have small motor units with each motor neuron controlling only a few muscle fibers. This allows very precise control, as activating one motor unit produces only a small, controlled contraction. Muscles that produce powerful but less precise movements like those in the legs and back have large motor units with each motor neuron controlling hundreds of muscle fibers. Activating one motor unit in these muscles produces stronger contraction but less precise control.

The brain controls the force of muscle contraction by adjusting how many motor units are activated and how rapidly they fire. Light contractions involve activating only a few motor units at low firing rates. Stronger contractions require recruiting more motor units and increasing their firing rates. This recruitment principle allows smooth gradation of force from very gentle to very powerful contractions using motor units that individually produce all-or-nothing responses.

The Autonomic Nervous System: The Unconscious Controller

While the somatic nervous system handles conscious sensation and voluntary movement, the autonomic nervous system regulates internal organs and physiological processes that normally occur without conscious control or awareness. The autonomic nervous system controls heart rate, blood pressure, digestion, breathing, pupil size, glandular secretions, and numerous other functions essential for survival. This system operates largely automatically, adjusting physiological parameters to maintain homeostasis and respond to changing demands.

The autonomic nervous system divides into three branches: the sympathetic nervous system, the parasympathetic nervous system, and the enteric nervous system. The sympathetic and parasympathetic systems often have opposing effects on target organs, with the balance between them determining overall activity level. This arrangement allows precise regulation, much like controlling room temperature by adjusting both heating and cooling systems.

The sympathetic nervous system mobilizes the body's resources for action, producing the constellation of physiological changes sometimes called the fight-or-flight response. When you encounter something threatening or stressful, sympathetic activation increases heart rate and blood pressure, dilates airways in the lungs, releases stored glucose for energy, redirects blood flow from digestive organs to skeletal muscles, dilates pupils, and triggers the release of stress hormones including adrenaline from the adrenal glands. These changes prepare your body for physical action, whether fighting a threat or fleeing from danger.

Sympathetic neurons originate in the middle regions of the spinal cord, specifically the thoracic and lumbar segments. Their axons exit the spinal cord and connect to a chain of ganglia, clusters of neuron cell bodies, that run parallel to the spinal cord. Most sympathetic neurons release norepinephrine onto target organs, though some release acetylcholine. The sympathetic system also includes specialized neurons in the adrenal medulla, the inner portion of the adrenal glands, which release adrenaline directly into the bloodstream when activated. This hormonal release amplifies and prolongs sympathetic effects, ensuring that the body remains mobilized until the threat passes.

The parasympathetic nervous system promotes rest, digestion, and recovery, sometimes called the rest-and-digest response. Parasympathetic activation slows heart rate, lowers blood pressure, constricts pupils, stimulates digestion and nutrient absorption, and generally promotes physiological states conducive to rest and recovery. While sympathetic activation prepares you for emergency action, parasympathetic activation allows your body to conserve energy and perform maintenance functions like digestion and tissue repair.

Parasympathetic neurons originate in the brain stem and the sacral, or lowest, region of the spinal cord. Unlike the sympathetic system with its chain of ganglia near the spinal cord, parasympathetic ganglia are located close to or within the organs they control. Parasympathetic neurons release acetylcholine onto target organs. The vagus nerve, the tenth cranial nerve originating in the brain stem, carries most parasympathetic signals to organs in the chest and abdomen, including the heart, lungs, stomach, and intestines. This single nerve exerts enormous influence over internal physiology, slowing the heart, promoting digestion, and coordinating many aspects of the calm, relaxed physiological state.

Most internal organs receive input from both sympathetic and parasympathetic systems, allowing precise regulation through the balance between excitation and inhibition. The heart provides a clear example. Sympathetic input increases heart rate and contraction strength, while parasympathetic input through the vagus nerve decreases heart rate and contraction strength. Your heart rate at any moment reflects the balance between these opposing influences. During exercise or stress, sympathetic activity dominates, increasing heart rate to deliver more oxygen to working muscles. During rest and relaxation, parasympathetic activity dominates, slowing heart rate to conserve energy. This dual control allows rapid, precise adjustments in heart rate to match changing demands.

The enteric nervous system consists of an extensive network of neurons embedded in the walls of the gastrointestinal tract, from the esophagus to the anus. This system is sometimes called the second brain because it contains more neurons than the spinal cord and can control digestive processes independently of the brain and spinal cord. The enteric nervous system regulates gut motility, secretion of digestive enzymes and fluids, and blood flow in the intestinal walls. While the enteric system can function autonomously, it also receives input from sympathetic and parasympathetic systems that modulate its activity based on overall physiological state.

Recent research has revealed surprising connections between the enteric nervous system and brain function. The gut produces many of the same neurotransmitters found in the brain, including serotonin, dopamine, and GABA. In fact, the gut produces more serotonin than the brain does, though this serotonin cannot cross the blood-brain barrier and therefore does not directly affect brain serotonin levels. Nevertheless, the gut communicates with the brain through the vagus nerve and through hormones and immune signals, creating what is called the gut-brain axis. This communication appears to influence mood, stress responses, and even cognitive function, though the mechanisms remain incompletely understood.

The relationship between gut and brain may explain why digestive problems often accompany stress and anxiety, and why some gastrointestinal disorders are associated with depression and anxiety disorders. Irritable bowel syndrome, characterized by abdominal pain and altered bowel habits, shows high rates of comorbidity with anxiety and depression. Whether gut dysfunction contributes to psychological symptoms, psychological symptoms contribute to gut dysfunction, or both influence each other through bidirectional communication remains an active area of research with important implications for treating both gastrointestinal and psychiatric disorders.

The Stress Response: Preparing for Threat

The autonomic nervous system's most dramatic function is orchestrating the body's response to stress and threat. This response, refined over millions of years of evolution, prepares organisms to deal with immediate dangers through a coordinated set of physiological changes. Understanding the stress response reveals how psychological states like fear and anxiety are inseparable from bodily states, and how chronic stress can damage both physical and mental health.

When you perceive something threatening, information reaches the amygdala, the brain structure discussed in earlier chapters that serves as a threat detector. The amygdala rapidly activates two major stress response systems: the sympathetic nervous system for immediate response and the hypothalami-pituitary-adrenal axis for sustained response. These systems work together to mobilize energy, focus attention, and prepare the body for action.

The sympathetic nervous system responds within seconds of threat detection. Sympathetic neurons increase heart rate and blood pressure, ensuring that muscles receive adequate oxygen and nutrients. Airways dilate, allowing more air into the lungs. Pupils dilate, improving vision. Blood flow redirects from digestive organs and skin toward skeletal muscles that might be needed for fighting or fleeing. The liver releases stored glucose into the bloodstream, providing quick energy. Sweat production increases, helping dissipate heat generated by physical activity. The adrenal medulla releases adrenaline and norepinephrine into the bloodstream, amplifying and prolonging these effects throughout the body.

These physiological changes create the subjective sensations associated with fear and anxiety. Your heart pounds. Your breathing quickens. Your muscles tense. You might feel butterflies in your stomach as blood flow shifts away from digestive organs. Your hands might become cold and sweaty as blood flow redirects and sweat glands activate. You might feel a surge of energy and heightened alertness. These sensations are not side effects of fear but integral components of it. The psychologist William James famously argued that we do not run because we are afraid; rather, we are afraid because we run. While James overstated the case, his insight that emotion involves bodily changes as much as mental states has been confirmed by modern neuroscience.

The hypothalami-pituitary-adrenal axis, commonly called the HPA axis, provides a slower but more sustained stress response. When the amygdala detects threat, it activates the hypothalamus, which releases corticotropin-releasing hormone. This hormone travels the short distance to the pituitary gland, stimulating it to release adrenocorticotropic hormone into the bloodstream. This hormone circulates to the adrenal glands, specifically the adrenal cortex or outer portion, causing release of cortisol, the primary stress hormone.

Cortisol mobilizes energy stores by promoting the breakdown of proteins and fats into glucose. It enhances the effects of adrenaline on the cardiovascular system. It suppresses nonessential functions like digestion, growth, reproduction, and immune responses, redirecting resources toward immediate survival. Cortisol also affects the brain, influencing memory formation, attention, and mood. Moderate cortisol levels can enhance memory for stressful events, ensuring you remember dangerous situations to avoid them in the future. However, excessive or prolonged cortisol exposure can impair memory and damage brain structures, particularly the hippocampus.

The HPA axis includes negative feedback mechanisms that turn off the stress response once threat passes. Cortisol binds to receptors in the hypothalamus and pituitary gland, inhibiting further release of the hormones that stimulate cortisol production. This feedback loop prevents excessive stress hormone levels and allows the body to return to baseline once the stressor is removed. The parasympathetic nervous system also becomes active during recovery, slowing heart rate, promoting digestion, and restoring the calm physiological state appropriate when danger has passed.

This stress response system evolved to handle acute physical threats that require immediate action and then quickly resolve. A gazelle fleeing from a lion experiences massive stress system activation during the chase, but if it escapes, the stress response rapidly shuts down and the gazelle returns to peacefully grazing. The system works beautifully for short-term stressors followed by recovery periods.

However, modern human life often involves chronic psychological stressors that activate the stress response repeatedly or continuously without opportunity for full recovery. Work pressures, financial worries, relationship conflicts, and constant information about threats near and far can keep stress systems chronically activated. Unlike acute physical threats, these psychological stressors cannot be resolved through fighting or fleeing. Your body mobilizes for action, but the action cannot occur, leaving you in a state of prolonged physiological arousal.

Chronic stress activation has devastating effects on physical and mental health. Sustained elevations in blood pressure damage blood vessels and increase risk of heart disease and stroke. Suppression of immune function increases vulnerability to infections and may contribute to cancer risk. Chronic cortisol elevation can cause muscle wasting, bone loss, and fat accumulation, particularly abdominal fat that carries especially high health risks. In the brain, chronic stress shrinks the hippocampus, impairs memory, and increases risk of depression and anxiety disorders. Chronic stress also disrupts sleep, appetite regulation, and numerous other physiological processes, creating a vicious cycle where stress causes problems that create more stress.

The damage caused by chronic stress explains why psychological factors have profound effects on physical health. People experiencing chronic stress show higher rates of cardiovascular disease, diabetes, gastrointestinal problems, chronic pain, and premature death. The pioneering research of physiologist Hans Selye in the mid-twentieth century demonstrated that sustained stress produces a general adaptation syndrome with predictable stages of alarm, resistance, and eventually exhaustion as the body's ability to cope becomes overwhelmed. Selye's work established that psychological stress is not merely subjective discomfort but a biological state with measurable physiological consequences.

Understanding the stress response has important implications for managing stress and protecting health. Interventions that reduce stress system activation or promote parasympathetic recovery can buffer stress effects. Regular exercise reduces stress hormone levels and improves stress resilience. Meditation and relaxation techniques activate parasympathetic responses and reduce sympathetic arousal. Social support appears to buffer stress effects through multiple mechanisms including reduced stress hormone release and enhanced immune function. Even simple practices like deep breathing can shift autonomic balance toward parasympathetic dominance, creating immediate physiological calm.

Cognitive approaches to stress management work partly by changing how situations are appraised. The stress response depends not just on objective threat but on whether something is perceived as threatening or controllable. Two people experiencing the same objectively stressful situation may have very different stress responses depending on how they interpret and respond to it. Someone who views a challenging situation as an opportunity may show minimal stress activation, while someone who views it as an overwhelming threat may show massive activation. Cognitive therapy helps people reappraise stressors in ways that reduce perceived threat and promote adaptive coping, thereby reducing stress system activation and its harmful effects.

Measuring Autonomic Function: Windows into Physiological State

Because autonomic nervous system activity profoundly influences physiological state and emotional experience, researchers have developed numerous methods to measure autonomic function. These measurements provide objective indices of stress, emotion, attention, and arousal that complement subjective self-reports.

Heart rate variability, the variation in time intervals between heartbeats, provides a sensitive measure of autonomic balance. High heart rate variability, with substantial beat-to-beat fluctuation, indicates strong parasympathetic influence and good autonomic regulation. Low heart rate variability, with little fluctuation, suggests either high sympathetic activation or poor autonomic regulation and is associated with stress, anxiety, depression, and cardiovascular disease. Heart rate variability can be measured non-invasively and has become increasingly accessible through commercial fitness devices, making it a practical tool for monitoring stress and recovery.

Electrodermal activity, also called skin conductance or galvanic skin response, measures electrical conductance of the skin, which increases when sweat glands become active. Because sweat gland activity is controlled exclusively by the sympathetic nervous system and responds to both physical and psychological arousal, electrodermal activity provides a direct measure of sympathetic activation. This measure is used extensively in emotion research and is one component of traditional polygraph or lie detector tests, though its validity for detecting deception is questionable.

Pupil diameter changes with autonomic activity, dilating with sympathetic activation and constricting with parasympathetic activation. Modern eye-tracking technology can measure tiny pupil changes that reflect cognitive effort, emotional arousal, and attention. Researchers use pupil measurements to study mental workload, decision-making processes, and emotional responses with millisecond precision.

Salivary cortisol provides a non-invasive way to measure HPA axis activity. Because cortisol levels in saliva closely track blood cortisol levels, repeated saliva samples throughout the day can reveal patterns of cortisol secretion and responses to stressors. This method has revealed that cortisol typically follows a daily rhythm, highest shortly after waking and declining throughout the day. Disruptions in this rhythm are associated with chronic stress, depression, and various health problems.

These physiological measures reveal that autonomic responses often occur before conscious awareness of emotion. When people view emotional images, their skin conductance and pupil diameter change within a few hundred milliseconds, well before they consciously recognize the emotion elicited by the image. This suggests that bodily responses are not simply reactions to consciously experienced emotions but rather integral components of emotional processing that can occur outside awareness.

Autonomic Disorders: When Control Systems Fail

Dysfunction of the autonomic nervous system produces diverse and often debilitating symptoms that dramatically impact quality of life. These disorders reveal how dependent we are on normally functioning autonomic control for basic activities like standing, eating, and maintaining stable internal physiology.

Orthostatic intolerance describes conditions where standing upright causes symptoms including dizziness, lightheadedness, rapid heartbeat, weakness, and sometimes fainting. When you stand up, gravity pulls blood toward your legs, potentially reducing blood flow to the brain. The autonomic nervous system normally compensates immediately by constricting blood vessels and increasing heart rate, maintaining adequate brain blood flow. People with autonomic dysfunction may lack adequate compensatory responses, allowing blood pressure to drop when standing. Severe cases may require medications that increase blood pressure or blood volume, wearing compression garments to prevent blood pooling in the legs, or even restricting activity to avoid symptoms.

Dysautonomia is a general term for conditions involving autonomic nervous system dysfunction. Causes include genetic disorders, autoimmune diseases, diabetes, Parkinson's disease, and other conditions that damage autonomic nerves. Symptoms are highly variable but can include abnormal heart rate or blood pressure, digestive problems, bladder dysfunction, temperature regulation difficulties, and abnormal sweating. Because the autonomic system controls so many different functions, dysautonomia can produce seemingly unrelated symptoms affecting multiple body systems, sometimes leading to misdiagnosis or dismissal of symptoms as psychological rather than physiological.

Postural orthostatic tachycardia syndrome, abbreviated POTAS, is a form of dysautonomia characterized by excessive heart rate increases upon standing, typically defined as an increase of thirty beats per minute or more. Patients experience debilitating symptoms when upright but feel better when lying down. The condition disproportionately affects young women and is often misdiagnosed as anxiety because rapid heartbeat and dizziness are common anxiety symptoms. However, in POTAS these symptoms result from physiological autonomic dysfunction rather than psychological anxiety, though the two can coexist and exacerbate each other.

Understanding these autonomic disorders has improved dramatically in recent years, leading to better diagnosis and treatment. Specialized autonomic testing can assess how well the autonomic nervous system responds to challenges like standing, deep breathing, or temperature changes. These objective tests help distinguish autonomic disorders from anxiety disorders and guide treatment selection. Increased awareness has also reduced the tendency to dismiss autonomic symptoms as psychosomatic, validating patients' experiences and directing them toward appropriate care.

Integration: Body and Mind as One System

This exploration of the nervous system reveals that the traditional separation between mind and body is artificial. Your thoughts and emotions depend on neural activity in your brain, but your brain depends on information from your body and controls physiological states throughout your body. You cannot have psychological experiences without bodily states, and bodily states profoundly influence psychological experiences.

The nervous system creates an integrated whole where perception depends on sensory neurons gathering information, where action depends on motor neurons controlling muscles, where emotions depend on autonomic changes in heart rate and hormones, and where all these processes are coordinated by the brain and spinal cord working together. Understanding this integration reveals why purely psychological interventions can affect physical health, why physical illnesses can produce psychological symptoms, and why effective healthcare must address both mind and body.

5

WHAT ARE EMOTIONS? A BIOLOGICAL PERSPECTIVE

Emotions color every aspect of human experience. They motivate your actions, guide your decisions, shape your relationships, and give meaning to your life. Joy makes achievements worth pursuing. Fear protects you from danger. Anger mobilizes you to defend against threats. Sadness signals loss and prompts withdrawal and reflection. Love binds you to others and motivates care and sacrifice. Disgust protects you from contamination and disease. These powerful experiences feel deeply personal and subjective, yet they arise from biological processes that have been shaped by millions of years of evolution.

For much of human history, emotions were viewed with suspicion by philosophers and scientists who valued reason and rationality. Emotions were seen as primitive impulses that interfered with logical thinking, as weaknesses to be overcome through willpower and discipline. The ancient Stoic philosophers taught that the ideal life involved freedom from emotional disturbance through rational control. Enlightenment thinkers championed reason over passion. Even early psychologists often treated emotions as disruptive forces that needed to be suppressed or redirected.

Modern neuroscience tells a very different story. Far from being irrational disruptions, emotions are sophisticated information-processing systems that evolved because they promoted survival and reproduction. Emotions provide rapid assessments of situations, preparing your body and mind to respond appropriately. They communicate information to others through facial expressions, body language, and vocalizations. They create memories that help you learn from experience. They motivate behavior in ways that conscious reasoning alone cannot. People with damage to brain emotion systems often make terrible decisions despite having intact reasoning abilities, demonstrating that good judgment requires integrating emotional information with logical analysis.

This chapter explores emotions from a biological perspective, examining the brain structures that generate emotional experiences, the bodily changes that accompany emotions, the evolutionary origins of different emotions, and how emotions are expressed and recognized across cultures. We will see that emotions are not arbitrary subjective feelings but rather organized patterns of neural, physiological, and behavioral responses that have been refined through evolution to help organisms navigate complex environments and social relationships.

Defining Emotion: More Than Just Feelings

Before examining the biology of emotions, we need to clarify what emotions are. In everyday language, people use words like emotion, feeling, mood, and affect somewhat interchangeably, but psychologists and neuroscientists draw important distinctions among these terms.

An emotion is a complex psychological state that involves three distinct components: a subjective experience or feeling, a physiological response involving changes throughout the body, and a behavioral or expressive response. When you experience fear, for example, you have the subjective feeling of being afraid, your body undergoes changes including increased heart rate and muscle tension, and you may display fearful facial expressions or engage in avoidant behavior. All three components typically occur together, though they can be dissociated under certain circumstances.

A feeling refers specifically to the subjective experiential component of emotion, the conscious awareness of an emotional state. You can have physiological arousal and behavioral responses without necessarily being consciously aware of feeling a particular emotion, particularly when responses occur very rapidly or outside attention. Conversely, you might feel an emotion without showing obvious behavioral signs if you deliberately suppress emotional expression.

A mood is a more diffuse and longer-lasting emotional state that is less intense than an emotion and not necessarily directed at a specific object or triggered by a particular event. While emotions are typically brief and have clear triggers, moods can last for hours or days and may not have obvious causes. You might wake up in an irritable mood without knowing why, and this mood colors your interpretations and reactions throughout the day. Depression and anxiety are sometimes described as mood disorders because they involve sustained alterations in emotional state rather than just intense emotional reactions to specific situations.

Affect is a general term encompassing all evaluative feelings, including emotions, moods, and even very simple positive or negative reactions. When researchers refer to positive affect and negative affect, they mean the general tendency to experience pleasant versus unpleasant emotional states. Some people are temperamentally high in positive affect, experiencing frequent joy, enthusiasm, and contentment. Others are high in negative affect, experiencing frequent anxiety, sadness, and anger. These tendencies appear to have genetic components and remain relatively stable throughout life, though they can be influenced by life circumstances and deliberate efforts to change habitual patterns of thought and behavior.

Emotions serve multiple functions that promoted survival and reproduction throughout evolutionary history. At the most basic level, emotions provide rapid evaluations of situations, automatically categorizing events as good or bad, safe or dangerous, approachable or avoidable. This automatic evaluation happens faster than conscious deliberation and often influences behavior before you fully understand why you feel a certain way. When you encounter a snake, fear arises immediately, prompting you to jump back before you consciously identify the snake or decide how to respond. This rapid response system has obvious survival value when dealing with immediate threats.

Emotions also prepare the body for action through coordinated physiological changes. Fear activates the sympathetic nervous system, increasing heart rate and blood pressure, releasing glucose for energy, and preparing muscles for rapid movement. Anger produces similar physiological arousal but with different action tendencies, preparing you to confront rather than flee from threats. Disgust triggers nausea and the urge to withdraw from contaminated objects or situations, protecting against ingesting harmful substances or contracting diseases. These physiological preparations occur automatically and coordinate multiple body systems to support appropriate behavioral responses.

Beyond their individual functions, emotions serve crucial social functions. Facial expressions, vocal tones, and body language communicate emotional states to others, allowing coordination of behavior and rapid transmission of important information. A fearful expression warns others of potential danger. An angry expression communicates willingness to defend resources or status. A sad expression elicits care and support from others. A joyful expression encourages approach and affiliation. These communicative functions make emotions essential for social living, allowing complex coordination without requiring explicit verbal communication.

Emotions also create memories and guide learning. Emotionally significant events are remembered better than neutral events, ensuring that important experiences leave lasting impressions that can guide future behavior. The fear you felt when you nearly had a car accident makes you more cautious in similar situations. The joy you experienced during a particular activity motivates you to repeat it. The embarrassment you felt after a social mistake helps you avoid similar mistakes in the future. This emotional tagging of memories prioritizes information storage, devoting limited memory resources to events most likely to be important for survival and well being.

The Limbic System Revisited: The Emotional Brain

As discussed in Chapter Two, a collection of brain structures collectively called the limbic system plays central roles in generating and regulating emotions. These structures, which include the amygdala, hippocampus, cingulate cortex, and several other regions, are evolutionarily ancient, found in similar forms across many mammalian species. This evolutionary conservation suggests that the basic mechanisms of emotion are shared across species, though humans also have uniquely developed cortical regions that allow more sophisticated emotional regulation and awareness.

The amygdala sits at the center of the brain's emotion systems, particularly for processing fear and other negative emotions. This almond-shaped structure receives inputs from all sensory modalities, allowing it to monitor the environment for potential threats. When the amygdala detects something threatening, it triggers a cascade of responses including activation of the sympathetic nervous system, release of stress hormones, enhanced attention and memory for the threatening stimulus, and defensive behaviors like freezing or fleeing.

Research using brain imaging has revealed that the amygdala responds not only to obvious threats but also to subtle emotional cues in facial expressions, tone of voice, and body language. When people view fearful faces, even when those faces are presented so briefly that viewers are not consciously aware of seeing them, the amygdala shows increased activity. This demonstrates that emotional processing can occur outside conscious awareness, with the amygdala automatically scanning for signs of potential danger.

The amygdala is not simply a fear center, as is sometimes suggested in oversimplified accounts. It also responds to positive stimuli, particularly those that are novel or salient, and plays important roles in learning associations between stimuli and emotional outcomes. Through a process called fear conditioning, the amygdala learns to associate previously neutral stimuli with aversive outcomes, creating learned fears. If you are bitten by a dog, your amygdala forms an association between dogs and pain, making you more fearful of dogs in the future. This learning happens rapidly, often requiring only a single pairing of stimulus and outcome, and the learned association can persist for years.

Fear conditioning demonstrates both the adaptive value and potential problems of emotional learning. The ability to learn fears quickly after a single bad experience protects you from repeating dangerous mistakes. However, this same mechanism can create persistent, irrational fears when the learned association is no longer appropriate. A person who develops a fear of flying after experiencing severe turbulence may continue fearing flights even though flying is statistically very safe. The amygdala has learned to associate planes with danger, and this learned association continues to trigger fear responses even when conscious reasoning says flying is safe.

The hippocampus, adjacent to the amygdala, provides contextual information that helps determine whether fear responses are appropriate. The hippocampus encodes where and when events occur, allowing you to distinguish safe from dangerous contexts. A combat veteran might learn through hippocampal processing that sudden loud noises in a war zone signal danger but similar noises in civilian life do not. When hippocampal function is impaired, this contextual discrimination breaks down, and fear responses may generalize inappropriately to safe situations. This may contribute to post-traumatic stress disorder, where trauma-related fears persist and intrude into daily life long after the original trauma has ended.

The prefrontal cortex, particularly the ventromedial and orbitofrontal regions discussed in Chapter Two, integrates emotional information with goals, social norms, and long-term consequences to guide decision-making. These regions receive inputs from the amygdala and other limbic structures and can modulate their activity, providing a mechanism for emotion regulation. When you feel angry but choose not to lash out because you recognize that doing so would damage an important relationship, prefrontal regions are suppressing the immediate emotional response in favor of long-term goals. When you feel anxious about giving a presentation but do it anyway because it is important for your career, prefrontal regions are overriding avoidance tendencies motivated by anxiety.

The capacity for emotion regulation depends heavily on prefrontal cortex function and develops slowly throughout childhood and adolescence as these brain regions mature. Young children have limited ability to regulate emotional responses, which is why they often have tantrums and emotional outbursts that seem disproportionate to adults. As the prefrontal cortex develops, children gradually gain better ability to control emotional impulses, delay gratification, and behave in ways that conflict with immediate emotional urges. This development continues well into the twenties, which helps explain why adolescents often show poor emotional control and engage in risky behaviors despite understanding potential consequences.

The anterior cingulate cortex, located in the medial surface of the frontal lobes, plays important roles in detecting conflicts between different response tendencies and signaling the need for increased cognitive control. When you experience conflicting emotions or when your emotional impulses conflict with your goals, the anterior cingulate becomes active and recruits prefrontal regions to resolve the conflict. This structure also appears to be involved in the emotional aspects of pain and in the distress caused by social rejection, suggesting that social and physical pain may share common neural mechanisms.

The insula, a region of cortex folded deep within the lateral fissure separating the temporal and frontal lobes, integrates information about internal bodily states with emotional experience. The insula receives input from sensory receptors throughout the body that monitor physiological conditions including heart rate, breathing, gut sensations, and muscle tension. This information contributes to the subjective feeling of emotions, to awareness of bodily states, and to feelings like disgust, craving, and pain. People with damage to the insula may have difficulty recognizing their own emotions or distinguishing different emotional states, suggesting this structure is crucial for the subjective feeling component of emotion.

These brain structures do not work in isolation but rather form interconnected circuits that generate coordinated emotional responses. When you encounter something emotionally significant, information flows rapidly through these circuits, with the amygdala providing initial threat detection, the hippocampus supplying contextual information, the insula contributing bodily state information, and the prefrontal cortex integrating all this information with goals and social considerations to guide behavior. Understanding emotions requires understanding not individual structures but the circuits and interactions among them.

The Physiology of Emotion: How Feelings Manifest in the Body

Emotions are not purely mental phenomena occurring only in the brain. Every emotion involves coordinated changes throughout the body, including alterations in heart rate, blood pressure, breathing, muscle tension, hormone levels, immune function, and numerous other physiological parameters. These bodily changes are not mere side effects of emotions but integral components that contribute to emotional experience and prepare the body for action.

Different emotions produce distinct patterns of physiological arousal. Fear and anger both activate the sympathetic nervous system and increase arousal, but with subtle differences. Fear tends to produce greater increases in certain stress hormones and stronger activation of brain regions involved in threat detection, while anger produces stronger activation of regions involved in approach motivation. Joy and contentment activate different patterns altogether, often involving parasympathetic activation and release of endorphins and other neurotransmitters associated with pleasure and reward.

Early emotion researchers debated whether these physiological patterns were specific enough to distinguish different emotions or whether all emotions involved similar generalized arousal that only became a specific emotion through cognitive interpretation. The James-Lange theory of emotion, proposed independently by William James and Carl Lange in the late nineteenth century, suggested that emotions arise from perception of bodily changes. According to this view, you do not run because you are afraid; rather, you perceive yourself running and interpret this as fear. You do not strike someone because you are angry; rather, you perceive your aggressive actions and interpret this as anger.

The competing Cannon-Bard theory, proposed by Walter Cannon and Philip Bard in the early twentieth century, argued that physiological arousal and emotional experience occur simultaneously and independently rather than arousal causing emotion. Cannon noted that physiological responses are too slow and too similar across different emotions to fully account for the speed and specificity of emotional experience. He also noted that people with spinal cord injuries who lack sensation from most of their body still experience emotions, suggesting that feedback from bodily changes is not necessary for emotional experience.

Modern research suggests that both theories capture important aspects of emotion. Physiological responses do differ somewhat across emotions, though perhaps not as dramatically as the James-Lange theory suggested. People can identify emotions somewhat accurately from physiological patterns alone, though not perfectly. At the same time, cognitive interpretation clearly influences emotional experience, with the same pattern of physiological arousal potentially being experienced as different emotions depending on context and interpretation, as the Cannon-Bard theory emphasized.

A famous experiment by psychologists Stanley Schachter and Jerome Singer demonstrated the role of cognitive interpretation in emotion. They gave participants injections of adrenaline, which causes physiological arousal including increased heart rate and trembling, but gave different participants different information about what to expect. Some were accurately told they would experience arousal. Others were told the injection would have no effects. Participants were then placed in situations designed to elicit either euphoria or anger. Those who were not warned about the arousal attributed it to the emotional situation and reported experiencing stronger emotions than those who correctly attributed the arousal to the drug. This demonstrated that unexplained physiological arousal can be interpreted as emotion depending on situational context.

This finding has important implications. It suggests that physiological arousal from one source can be misattributed to another source, intensifying emotional experiences. If you are physiologically aroused from exercise and then encounter an attractive person, you might misattribute some of your exercise-induced arousal to attraction, experiencing stronger romantic feelings than you would have without the exercise. If you are aroused from stress at work and then encounter a minor annoyance at home, you might misattribute your arousal to the annoyance and respond with disproportionate anger. This misattribution effect demonstrates that emotional experience depends on both physiological state and cognitive interpretation of that state.

The facial feedback hypothesis proposes another way that bodily states influence emotional experience. According to this hypothesis, facial expressions do not merely reflect emotions but can actually influence the emotions you experience. When you smile, even if you do not initially feel happy, the act of smiling sends feedback to your brain that can create or enhance positive feelings. When you frown, you may feel somewhat sadder. When you adopt an angry facial expression, you may feel somewhat angrier. This effect has been demonstrated in numerous studies, though the size of the effect and the mechanisms involved remain debated.

The facial feedback effect has practical implications. It suggests that deliberately adopting positive facial expressions might improve your mood, while chronically adopting negative expressions might worsen it. Some research suggests that people who receive botulinum toxin injections that paralyze facial muscles report reduced emotional intensity, possibly because they cannot produce the facial expressions that normally provide feedback contributing to emotional experience. However, other studies have failed to replicate this effect, and the relationship between facial expressions and emotional experience remains an active area of research and debate.

Beyond facial expressions, body posture and movement also influence emotional experience. Standing in an expansive, confident posture makes people feel more powerful and confident. Slumping in a withdrawn, defeated posture makes people feel less powerful and more stressed. Walking with an upright, bouncy gait creates more positive mood than walking with a slouched, shuffling gait. These effects demonstrate that the relationship between mind and body is bidirectional, with mental states influencing bodily states and bodily states influencing mental states.

Universal Emotions: Evidence from Evolution and Culture

One of the most fascinating questions about emotions is whether they are universal across cultures or culturally specific. Do people everywhere experience and express emotions in similar ways, or do emotions vary as much as languages and customs vary? The answer appears to be both. Certain basic emotions show remarkable universality, suggesting they are products of our shared evolutionary heritage, while other aspects of emotional experience vary across cultures, reflecting learning and social construction.

Psychologist Paul Ekman conducted groundbreaking research on the universality of facial expressions in the 1960s and 1970s. He traveled to remote, preliterate cultures in New Guinea that had minimal contact with Western societies and asked people to identify emotions from photographs of facial expressions. Despite enormous cultural differences, people in these isolated communities recognized happiness, sadness, anger, fear, disgust, and surprise from facial expressions with accuracy far above chance. Similarly, when Ekman asked these isolated people to make faces showing different emotions, Westerners could recognize the intended emotions with good accuracy.

These findings suggested that at least some emotions and their expressions are universal, not culturally learned. Ekman argued for a set of basic emotions, including happiness, sadness, anger, fear, disgust, and surprise, that are innate, universally recognized, and associated with distinct facial expressions and physiological patterns. These basic emotions presumably evolved because they promoted survival and reproduction, with each emotion solving specific adaptive problems.

Happiness signals that things are going well and motivates continued pursuit of whatever is producing positive outcomes. The facial expression of happiness, characterized by raised cheeks and corners of the mouth forming a smile, along with crow's feet wrinkles around the eyes in genuine smiles, communicates friendliness and approachability, facilitating social bonds.

Sadness signals loss or failure and motivates withdrawal and reflection. The facial expression of sadness, characterized by drooping eyelids and downturned mouth corners, along with reduced activity and energy, elicits care and support from others and provides time for recovery and reassessment of goals.

Anger arises when goals are blocked or when someone violates social rules or harms you. It motivates confrontation and aggressive defense of interests. The facial expression of anger, characterized by lowered brows, narrowed eyes, and tensed lips, along with physiological preparation for action, communicates determination and warns others not to interfere.

Fear arises in response to threat and motivates protective action including fleeing, freezing, or defensive aggression. The facial expression of fear, characterized by raised eyebrows, widened eyes, and open mouth, along with sympathetic nervous system activation, helps you detect and respond to danger while warning others of potential threats.

Disgust originally evolved to protect against ingesting harmful substances, with the facial expression involving nose wrinkling, raised upper lip, and tongue protrusion that helps expel contaminated material from the mouth. Disgust has expanded beyond food to include moral disgust at violations of social norms and values, suggesting that social and physical disgust may share common mechanisms.

Surprise arises in response to unexpected events and involves widened eyes and raised eyebrows that increase sensory intake, along with a cognitive reset that interrupts ongoing activity and directs attention to the unexpected stimulus.

While these basic emotions show substantial universality, many aspects of emotional experience and expression vary across cultures. Display rules, culturally learned norms about when and how emotions should be expressed, vary considerably. Some cultures encourage open emotional expression, while others value emotional restraint. Some cultures emphasize individual emotional experiences, while others emphasize collective emotional harmony. These cultural differences shape how people express, regulate, and even experience emotions.

More complex emotions like pride, shame, guilt, and embarrassment, sometimes called self-conscious emotions because they involve self-evaluation, show greater cultural variation. What situations elicit these emotions, how intensely they are experienced, and how they are expressed all vary across cultures depending on cultural values and social structures. In individualistic cultures that emphasize personal achievement, people may feel pride and shame based primarily on personal accomplishments and failures. In collectivistic cultures that emphasize social harmony and group belonging, people may feel pride and shame based on whether their actions honor or dishonor their family or group.

The universality of basic emotions suggests they are biological adaptations shaped by evolution, while cultural variation in complex emotions and emotional expression demonstrates the powerful influence of learning and culture. Emotions thus reflect both nature and nurture, with basic mechanisms shared across humanity and elaborations shaped by experience and culture.

Emotional Development: From Infancy to Adulthood

Emotions are present from birth, though emotional experience and regulation develop dramatically throughout childhood and adolescence. Even newborn infants display facial expressions recognizable as distress, disgust, and interest. Within the first few months of life, infants show expressions of joy, anger, and sadness. By the end of the first year, fear of strangers and separation anxiety emerge, reflecting developing attachments and awareness of unfamiliar people as potential threats.

More complex self-conscious emotions like embarrassment, pride, shame, and guilt emerge in the second and third years of life, once children develop self-awareness and begin to internalize social standards. These emotions require cognitive abilities including recognizing yourself as separate from others, holding standards for behavior, and evaluating whether your actions meet those standards. The emergence of self-conscious emotions coincides with major developmental milestones including language acquisition, theory of mind development, and increasing social awareness.

Throughout childhood and adolescence, children develop increasingly sophisticated abilities to understand, express, and regulate emotions. They learn to recognize and label a wider range of emotions in themselves and others. They develop better understanding of what causes different emotions and how emotions influence behavior. They learn cultural display rules about when emotional expression is appropriate and when it should be suppressed. They develop strategies for regulating emotions, progressing from relying on caregivers for comfort and regulation to developing internal regulatory capacities.

Emotion regulation, the ability to influence which emotions you experience, when you experience them, and how you express them, undergoes particularly dramatic development. Young children have limited regulatory abilities and depend heavily on caregivers to help them manage emotional distress. When a toddler is upset, parents provide soothing, distraction, and reassurance that help the child calm down. Through repeated experiences of being helped to regulate emotions, children gradually internalize these regulatory strategies and become able to soothe themselves.

As the prefrontal cortex develops throughout childhood and adolescence, regulatory abilities improve. Children become better able to delay gratification, suppressing the desire for immediate rewards in favor of larger delayed rewards. They become better able to control emotional impulses, thinking before acting when angry or frustrated. They develop more sophisticated regulatory strategies including reappraising situations to change how they feel, suppressing emotional expressions when socially appropriate, and seeking social support when overwhelmed.

However, emotion regulation abilities show significant individual differences, with some children showing much better regulation than others. These differences reflect both temperamental factors, biologically based tendencies toward emotional reactivity and self-control, and environmental factors, particularly the quality of early caregiving relationships. Children who experience sensitive, responsive caregiving develop better emotion regulation than children who experience inconsistent, harsh, or neglectful caregiving. Secure attachment relationships provide a safe base from which children can explore the world and learn to manage emotions, while insecure attachments are associated with heightened emotional reactivity and poorer regulation.

Adolescence brings new challenges for emotion regulation due to changes in brain development. The limbic system, including the amygdala and nucleus accumbens involved in emotional processing and reward, undergoes significant development during adolescence, becoming more reactive. Meanwhile, the prefrontal cortex, which provides regulatory control, continues developing well into the twenties. This imbalance between heightened emotional reactivity and still-developing regulatory control may contribute to the increased risk-taking, emotional volatility, and sensitivity to peer influence characteristic of adolescence.

Understanding emotional development has important implications for parenting and education. Helping children develop healthy emotion regulation requires both warmth and support during distress, and age-appropriate expectations and guidance that help children practice managing emotions. Punishing children for emotional expressions or expecting young children to regulate emotions beyond their developmental capacities can interfere with healthy emotional development. Conversely, always stepping in to manage children's emotions without allowing them opportunities to practice regulation also impairs development of self-regulatory skills.

Individual Differences in Emotional Experience

While all humans share basic emotional capacities, individuals differ substantially in their emotional experiences. Some people experience emotions intensely while others experience them mildly. Some people are generally happy and optimistic while others are anxious or melancholic. Some people express emotions freely while others are emotionally reserved. These individual differences have important consequences for wellbeing, relationships, and mental health.

Temperament refers to biologically based individual differences in emotional reactivity and regulation that are apparent from infancy and show some stability across the lifespan. Researchers have identified several dimensions of temperament including activity level, attention span, adaptability to new situations, intensity of emotional reactions, and general emotional tone. These temperamental dimensions show moderate heritability, meaning that genetic factors contribute to variation among individuals, though environmental factors also play important roles.

One influential model distinguishes between behavioral inhibition and behavioral approach systems. The behavioral inhibition system, associated with anxiety and negative emotions, responds to threats and punishment by promoting withdrawal and avoidance. The behavioral approach system, associated with positive emotions and reward seeking, responds to opportunities and rewards by promoting approach and engagement. People vary in the relative strength of these systems, with some showing strong inhibition and high anxiety, others showing strong approach and reward seeking, and still others showing balanced systems.

Neuroscience research has identified brain differences associated with these temperamental dimensions. People high in behavioral inhibition tend to show greater amygdala reactivity to threat-related stimuli and potentially altered connectivity between the amygdala and prefrontal regulatory regions. People high in behavioral approach show greater activity in dopamine-rich reward circuits when anticipating or experiencing rewards. These neural differences provide biological substrates for temperamental variation, though whether neural differences cause temperamental differences, result from them, or both remains incompletely understood.

Emotional intelligence, the ability to perceive, understand, and manage emotions in yourself and others, represents another important dimension of individual difference. People high in emotional intelligence are skilled at recognizing subtle emotional cues, understanding what causes different emotions, regulating their own emotional responses, and using emotional information to guide thinking and behavior. High emotional intelligence predicts better social relationships, superior job performance in positions requiring interpersonal skills, and better mental health outcomes.

Emotional intelligence appears to involve multiple abilities rather than a single unified skill. Some people excel at recognizing emotions in facial expressions but struggle to manage their own emotional reactions. Others are skilled at regulating emotions but less sensitive to others' emotional states. These different components of emotional intelligence likely depend on different neural systems, with emotion perception relying heavily on structures like the amygdala and superior temporal sulcus that process social stimuli, while emotion regulation relies more on prefrontal regions involved in cognitive control.

Understanding individual differences in emotional experience helps explain why people respond so differently to the same situations. What one person finds thrilling, another finds terrifying. What makes one person angry might not bother another person at all. These differences are not simply matters of choice or willpower but reflect underlying biological and psychological variation in how emotional systems function. Recognizing this variation can increase empathy and reduce judgment toward people whose emotional responses differ from our own.

Emotions and Decision-Making: The Wisdom of Feeling

For centuries, emotions and reason were viewed as opponents, with emotions seen as disruptions to rational thought that should be suppressed to make good decisions. Modern neuroscience tells a different story. Far from interfering with good judgment, emotions provide essential information that improves decision-making. People with damage to emotion-processing brain regions often make terrible choices despite having intact reasoning abilities, demonstrating that effective decision-making requires integrating emotional and logical information.

The most famous evidence comes from patients with damage to the ventromedial prefrontal cortex, a region that integrates emotional information with decision-making. The neurologist Antonio Damasio studied a patient known as Elliot who developed damage to this region following surgery to remove a brain tumor. After the surgery, Elliot showed normal intelligence and reasoning on standard tests. He could logically analyze situations and describe appropriate courses of action. However, in real life, he made disastrous decisions, losing his job, going bankrupt, and destroying his marriage through poor choices.

Damasio proposed that Elliot's problem was not too much emotion but too little. Without normal emotional responses to guide his decisions, Elliot could analyze options logically but could not sense which options felt right or wrong. He might spend hours deliberating over trivial choices like where to eat lunch because without emotional preferences to guide him, all options seemed equally valid logically. For important decisions, he could not use emotional reactions as signals of which choices aligned with his values and goals.

Damasio developed the somatic marker hypothesis to explain how emotions guide decision-making. According to this theory, when you contemplate different choices, your brain automatically generates emotional reactions based on past experiences with similar choices. These emotional reactions, which Damasio calls somatic markers because they involve bodily states, provide rapid signals about whether choices are likely to lead to good or bad outcomes. You might feel a sense of unease when considering a risky investment or a feeling of excitement when considering a new opportunity. These feelings incorporate learning from past experiences in a form that can guide decisions quickly without requiring conscious analysis.

Research using a gambling task developed by Damasio and colleagues demonstrates somatic markers in action. In this task, people choose cards from different decks, with some decks rigged to produce long-term losses despite occasional large wins, while other decks produce long-term gains through consistent small wins. Healthy participants typically develop advantageous strategies, avoiding the risky losing decks and favoring the safe winning decks. Importantly, they develop these preferences before they can consciously articulate which decks are better, and they show physiological stress responses, measured by skin conductance, when contemplating choices from the bad decks even before they consciously recognize these decks as risky. These unconscious emotional responses guide advantageous decision-making before conscious awareness catches up.

Patients with ventromedial prefrontal damage fail to develop these advantageous preferences. They continue choosing from the risky decks even after experiencing losses, and they do not show the anticipatory stress responses that normally guide healthy people away from bad choices. Without normal emotional signals, these patients cannot learn to avoid disadvantageous options despite understanding logically that some decks are worse than others.

These findings reveal that emotions are not irrational disruptions but sophisticated guidance systems that incorporate experiential learning to help you navigate complex environments. Emotions allow you to respond to situations based on accumulated experience without consciously remembering or analyzing every relevant past event. They provide rapid assessments when you do not have time for lengthy deliberation. They help you choose between options that are difficult to compare logically, such as deciding between different career paths or potential partners.

However, emotions can also lead to poor decisions when they trigger responses that were adaptive in ancestral environments but not in modern contexts. Fear of public speaking, for example, involves activation of threat-detection systems even though giving a presentation typically poses no actual danger. The same mechanisms that protect you from genuine physical threats can create anxiety in socially evaluative situations where the consequences are usually mild. Understanding when to trust emotional reactions and when to override them through reason is part of emotional wisdom.

The Biology of Emotional Expression and Recognition

Emotions are inherently social. They evolved not just to help individuals respond to situations but to coordinate behavior and communication within social groups. Emotional expressions provide rapid communication of internal states, allowing others to predict your likely behavior and adjust their own behavior accordingly. A fearful expression warns others of danger. An angry expression communicates that you are prepared to defend yourself. A happy expression invites approach and affiliation. This communicative function makes emotional expressions a crucial component of human social life.

Facial expressions are the most studied form of emotional expression, with distinctive patterns of facial muscle movements associated with different emotions. The smile of genuine happiness, called a Duchenne smile after the neurologist who first described it, involves not just raising the corners of the mouth but also contracting muscles around the eyes that create crow's feet wrinkles. This combination of muscle movements is difficult to produce voluntarily, making genuine smiles distinguishable from polite, socially motivated smiles that involve only the mouth. People unconsciously distinguish genuine from fake smiles, reacting more positively to genuine expressions.

The universality of facial expressions raises interesting questions about whether they are innate or learned. Several lines of evidence suggest substantial innate components. Children who are born blind and thus cannot learn expressions through imitation nonetheless produce recognizable emotional expressions, suggesting these expressions do not require visual learning. Different species of primates show similar expressions, suggesting these patterns are evolutionarily ancient. Human infants produce expressions that adults recognize before they could have learned appropriate expressions through feedback.

However, cultural learning clearly influences emotional expression. As mentioned earlier, cultures vary in display rules about when and how emotions should be expressed. People learn to amplify expressions in some contexts, suppress them in others, and sometimes display expressions they do not feel to conform to social expectations. Japanese participants in one study showed similar facial expressions to Americans when viewing emotional films alone, but when viewing with an authority figure present, Japanese participants more often masked negative expressions with smiles, reflecting cultural values emphasizing emotional restraint and respect for authority.

The brain has specialized systems for recognizing emotional expressions in others. Certain regions, particularly in the superior temporal sulcus and fusiform gyrus, respond preferentially to faces and show different patterns of activity when viewing different facial expressions. The amygdala becomes particularly active when viewing fearful faces, perhaps because fear expressions communicate potential threats in the environment. Damage to these face-processing regions can cause prosopagnosia, face blindness, where people cannot recognize individuals from their faces. More specific impairments affect emotion recognition while leaving face identity recognition intact, or vice versa, suggesting these functions depend on partially separate neural systems.

Mirror neurons, neurons that fire both when performing an action and when observing someone else perform that action, may contribute to understanding others' emotions. When you see someone smile, mirror neurons in your brain involved in smiling become active, potentially creating a faint echo of the observed emotion in your own system. This neural resonance might help you understand what the other person is feeling by creating a similar state in yourself. While mirror neuron research remains controversial and their precise functions debated, they provide one possible mechanism for empathy and emotional contagion.

Emotional contagion, the tendency to automatically mimic and synchronize with others' emotions, demonstrates the social nature of emotions. When you see someone yawning, you are more likely to yawn yourself. When you are around happy people, you tend to feel happier. When you are around anxious people, you tend to feel more anxious. This automatic sharing of emotional states helps coordinate group behavior and strengthens social bonds. However, it can also transmit stress and negative emotions through social networks, with implications for mental health and well being.

Understanding the biology of emotional expression and recognition has practical applications. Improving ability to recognize emotions in others might enhance social competence and relationship quality. Some research suggests that training in emotion recognition can improve social skills, particularly for individuals with autism spectrum disorders who often struggle to read emotional cues. Teaching people to attend to subtle facial expressions, body language, and vocal tones can enhance their ability to navigate social interactions successfully.

The rise of digital communication raises new questions about emotional expression and recognition. Text messages and emails lack the facial expressions, tone of voice, and body language that normally accompany emotional communication, creating frequent misunderstandings. People often interpret neutral messages as hostile or sarcastic when they were intended neutrally, demonstrating how much we depend on nonverbal emotional cues for accurate communication. Emojis and emoticons represent attempts to restore emotional information to text communication, providing simplified substitutes for facial expressions that can clarify intended emotional tone.

Video communication platforms partially restore nonverbal emotional cues, allowing people to see facial expressions and hear vocal tone. However, technical limitations including delays, frozen images, and restricted visual fields can disrupt the normal flow of emotional communication. Research on virtual communication during the COVID-19 pandemic revealed that many people found extensive video calls exhausting, a phenomenon dubbed Zoom fatigue. This exhaustion may partly reflect the cognitive effort required to decode emotional information under suboptimal conditions and the inability to use spatial proximity and body orientation cues that normally facilitate in-person interaction.

When Emotions Go Awry: Emotional Disorders

Given the complexity of emotional systems and their dependence on multiple brain regions, neurotransmitters, and hormones, it is perhaps unsurprising that many things can go wrong, producing the various emotional disorders that cause immense suffering. Understanding the biological basis of these disorders provides insight into their causes and guides development of more effective treatments.

Anxiety disorders involve excessive fear and worry that persist even in the absence of genuine threats. Generalized anxiety disorder is characterized by chronic worry about multiple domains of life that is difficult to control and causes significant distress or impairment. Panic disorder involves recurrent panic attacks, intense episodes of fear accompanied by physiological arousal including rapid heartbeat, sweating, trembling, and feelings of losing control or dying. Social anxiety disorder involves intense fear of social situations where one might be evaluated or embarrassed by others. Specific phobias involve intense fear of particular objects or situations such as heights, animals, blood, or enclosed spaces.

These disorders appear to involve dysfunction in the brain's threat detection and response systems. Neuroimaging studies consistently show heightened amygdala reactivity in anxiety disorders, suggesting that threat detection systems are oversensitive, responding to stimuli that are not actually dangerous. In some anxiety disorders, connectivity between the amygdala and prefrontal regulatory regions appears reduced, potentially impairing the ability to regulate anxiety responses once they are triggered. This neurobiological understanding helps explain why anxiety disorders are not simply matters of choosing to stop worrying, but rather reflect alterations in brain systems that operate largely outside voluntary control.

Neurotransmitter systems also play important roles in anxiety. The neurotransmitter GABA, which inhibits neural activity, appears to be underactive in many anxiety disorders. This reduced inhibition may allow fear circuits to become overactive. Anti-anxiety medications including benzodiazepines work by enhancing GABA activity, essentially increasing inhibition throughout the brain and thereby reducing anxiety. Serotonin also influences anxiety, and selective serotonin re uptake inhibitors are commonly prescribed for anxiety disorders as well as depression.

Depression, more formally called major depressive disorder, involves persistent sadness, loss of interest and pleasure in activities, changes in appetite and sleep, fatigue, difficulty concentrating, feelings of worthlessness or guilt, and sometimes thoughts of death or suicide. Depression is not simply feeling sad for a few days after a disappointment but rather a sustained alteration in mood and functioning that persists for at least two weeks and often much longer without treatment.

The biology of depression is complex and incompletely understood, involving multiple neurotransmitter systems, hormonal changes, alterations in brain structure, and disrupted neural circuits. The mono amine hypothesis, which dominated thinking about depression for decades, proposed that depression results from deficiencies in the mono amine neurotransmitters serotonin, norepinephrine, and dopamine. This hypothesis was based partly on observations that drugs increasing these neurotransmitters often alleviate depression, while drugs depleting them can cause depression.

However, the mono amine hypothesis is now recognized as oversimplified. Antidepressant medications increase mono amine levels within hours of first dose, yet their therapeutic effects typically require several weeks, suggesting they work through slower adaptive changes rather than simply increasing neurotransmitter levels. Moreover, some effective antidepressants do not primarily affect mono amine systems, and drugs that increase mono amines do not reliably improve mood in people without depression.

Current theories emphasize neural plasticity and circuit-level changes in depression. Depression is associated with reduced volume of the hippocampus and prefrontal cortex, potentially reflecting decreased neural plasticity and reduced formation of new connections. Chronic stress, a major risk factor for depression, causes similar structural changes through sustained elevation of cortisol and other stress mediators. Some researchers propose that antidepressants work partly by enhancing neural plasticity, promoting formation of new neural connections that help the brain escape from maladaptive patterns.

The brain circuits involved in reward processing and motivation also show altered function in depression. The ventral striatum, a key component of the brain's reward system, shows reduced activity in response to rewarding stimuli in depressed individuals. This reduced reward sensitivity may underlie anhedonia, the inability to experience pleasure that is a core symptom of depression. The prefrontal cortex, particularly regions involved in emotion regulation, also shows altered activity and connectivity in depression, potentially contributing to the inability to control negative thoughts and feelings.

Understanding depression as a disorder involving multiple neural systems rather than simply a chemical imbalance has led to development of diverse treatment approaches. In addition to medications that affect neurotransmitter systems, treatments now include psychotherapy approaches that target maladaptive thought patterns, behavioral activation that encourages engagement in rewarding activities, exercise programs that promote neural plasticity and improve reward system function, and even experimental treatments like ketamine that rapidly enhance neural plasticity through different mechanisms than traditional antidepressants.

Post-traumatic stress disorder, commonly abbreviated as PTSD, develops in some people following exposure to traumatic events involving actual or threatened death, serious injury, or sexual violence. PTSD is characterized by intrusive memories or flashbacks of the trauma, avoidance of reminders of the trauma, negative changes in thoughts and mood, and heightened arousal and reactivity. People with PTSD may feel constantly on edge, have difficulty sleeping, startle easily, and experience emotional numbness alternating with intense distress.

PTSD appears to involve dysfunction in multiple brain systems involved in fear, memory, and emotion regulation. The amygdala shows heightened reactivity, responding excessively to trauma reminders and potentially to neutral stimuli misperceived as threatening. The hippocampus, crucial for encoding contextual information about where and when events occur, often shows reduced volume in PTSD and may function abnormally. This hippocampal dysfunction may contribute to the over generalization of fear, where fear responses appropriate to the original trauma context intrude into safe contexts.

The prefrontal cortex, particularly the ventromedial region involved in fear extinction, shows reduced activity in PTSD. Fear extinction is the process by which learned fear responses diminish when the feared stimulus is repeatedly encountered without adverse consequences. Impaired extinction may explain why PTSD symptoms persist long after the trauma has ended, as the brain fails to update fear memories to reflect current safety. Treatments that enhance extinction, including exposure therapy where patients gradually confront trauma reminders in safe contexts, can reduce PTSD symptoms by helping the brain learn that trauma reminders are no longer dangerous.

The stress hormone system also shows alterations in PTSD, though these changes are complex and sometimes paradoxical. While acute stress causes cortisol elevation, chronic PTSD is sometimes associated with lower cortisol levels, possibly reflecting adaptation of the stress response system or altered feedback regulation. Some researchers have investigated whether administering cortisol shortly after trauma might prevent PTSD by strengthening the extinction of fear memories, though this approach remains experimental.

Bipolar disorder involves episodes of mania or hypomania, periods of abnormally elevated or irritable mood, increased energy, reduced need for sleep, racing thoughts, impulsivity, and sometimes psychosis, alternating with episodes of depression. The mood swings in bipolar disorder are much more extreme than normal mood variations and can cause severe impairment in functioning and relationships. Some people experience rapid cycling between manic and depressive states, while others have long periods of normal mood between episodes.

The biology of bipolar disorder remains poorly understood despite decades of research. The condition shows strong heritability, with genetic factors playing larger roles than in many other psychiatric disorders. However, no single gene causes bipolar disorder; rather, many genes with small individual effects combine with environmental factors to influence risk. Bipolar disorder involves dysfunction in circuits regulating mood, reward processing, and behavioral inhibition, though the specific alterations remain incompletely characterized.

Neurotransmitter systems clearly play roles, as medications that stabilize mood in bipolar disorder affect various neurotransmitter systems including glutamate, GABA, dopamine, and serotonin. Lithium, one of the oldest and most effective mood stabilizers, has complex effects on multiple cellular signaling pathways and may work partly by enhancing neural plasticity and protecting neurons from damage. Anticonvulsant medications originally developed for epilepsy also stabilize mood in bipolar disorder, possibly by enhancing inhibitory neurotransmission or reducing excessive neural excitability.

Brain imaging studies have found structural and functional differences in bipolar disorder, including alterations in prefrontal cortex, amygdala, and striatum. Some studies find reduced gray matter volume in prefrontal regions, potentially reflecting impaired development or accelerated loss of neurons. Others find altered connectivity between regions involved in emotion generation and regulation, potentially explaining the extreme mood swings characteristic of the disorder.

Understanding these emotional disorders as disorders of brain function rather than character flaws or moral failings reduces stigma and promotes more effective treatment. While environmental factors including stress, trauma, and adversity clearly influence risk for these disorders, their biological underpinnings explain why they cannot be overcome through willpower alone and why medical and psychological interventions targeting brain function are often necessary.

Measuring Emotions: Challenges and Methods

Studying emotions scientifically presents unique challenges because emotional experiences are inherently subjective and private. Researchers cannot directly observe what someone feels but must rely on indirect measures that each have limitations. Understanding these methods and their limitations is important for interpreting emotion research correctly.

Self-report measures, where people rate their emotional experiences using questionnaires or scales, are the most direct way to assess subjective emotional experience. People can report how happy, sad, anxious, or angry they feel, often using numerical scales or choosing from lists of emotion words. These reports provide valuable information about conscious emotional experience but have important limitations. People may not be fully aware of their emotions, particularly subtle or mixed emotions. Social desirability concerns may lead people to under report negative emotions or over report positive emotions. Memory for past emotions is often inaccurate, colored by current mood and by whether outcomes were good or bad.

Behavioral measures assess emotional expressions and actions rather than subjective experience. Researchers might code facial expressions using systems like the Facial Action Coding System, which describes facial muscle movements in detail, allowing objective measurement of emotional expressions. They might measure approach and avoidance behaviors, with approach indicating positive emotion and avoidance indicating negative emotion. Behavioral measures have the advantage of being objective and not requiring accurate introspection, but expressions can be controlled or suppressed, and the relationship between behavior and subjective experience is not always straightforward.

Physiological measures assess bodily changes associated with emotions, including heart rate, blood pressure, skin conductance, hormone levels, and immune function. These measures are objective and often reveal emotional arousal even when people deny feeling emotions. However, physiological arousal is often similar across different emotions, making it difficult to identify specific emotions from physiology alone. Moreover, physiological arousal can have multiple causes beyond emotion, requiring careful experimental control to interpret findings correctly.

Brain imaging measures, including functional MRI, PET scans, and EEG, allow researchers to observe brain activity associated with emotions. These techniques have revealed which brain regions are involved in different emotional processes and how patterns of brain activity differ between people with and without emotional disorders. However, current brain imaging technology has limitations in temporal and spatial resolution, making it difficult to capture the rapid, distributed neural processes underlying emotion. Moreover, brain activity patterns associated with emotions show substantial individual variation, making it difficult to decode specific emotions from brain activity alone.

Psycho physiological measures combine physiological recording with psychological manipulations to study specific emotional processes. For example, researchers might measure startle responses while people view emotional images. The startle reflex, typically measured by the eye blink response to a loud noise, is enhanced when people are viewing unpleasant images and reduced when viewing pleasant images, providing an objective measure of emotional valence that does not require conscious awareness or reporting.

Each measurement approach provides partial information about emotions, and the strongest conclusions come from studies that use multiple converging methods. If self-reports, behaviors, physiology, and brain activity all point to the same conclusion, confidence in that conclusion increases. When different measures disagree, this often reveals important complexities, such as conscious emotional experience differing from unconscious emotional responses.

The Future of Emotion Science

Research on the biology of emotions continues to advance rapidly, with new technologies providing unprecedented opportunities to understand how emotions are generated, regulated, and expressed. Improved brain imaging techniques with better spatial and temporal resolution are revealing the dynamics of emotional processing in real time. Genetic studies are identifying genes that influence individual differences in emotional reactivity and regulation. Animal research using optogenetics, which allows researchers to activate or silence specific neurons using light, is revealing causal relationships between neural activity and emotional behaviors.

These advances are translating into improved treatments for emotional disorders. Researchers are developing more targeted medications that affect specific components of emotional systems rather than broadly increasing or decreasing neurotransmitter levels. Psychotherapies are being refined based on understanding of the neural mechanisms they target. Novel interventions including neurofeedback, where people learn to control their own brain activity, and trans cranial magnetic stimulation, which can modulate activity in specific brain regions, offer new options for people who do not respond to traditional treatments.

However, important questions remain unanswered. How exactly does neural activity create subjective emotional experience? Why do some people develop emotional disorders after trauma while others remain resilient? How can treatments be better personalized to individual patients based on their specific biological profiles? What are the long-term consequences of manipulating emotional systems through medications or other interventions? Answering these questions will require continued research integrating multiple levels of analysis from genes to neural circuits to behavior.

Understanding emotions as biological phenomena does not diminish their importance or reduce their complexity. Rather, it reveals emotions as sophisticated systems that have been refined through millions of years of evolution to help organisms survive, thrive, and connect with others. Emotions are not primitive impulses to be suppressed but essential sources of information that should be understood and integrated with reason to guide wise action. The biology of emotions provides tools for understanding when emotional responses serve us well and when they lead us astray, empowering people to work with rather than against their emotional nature.

6

FEAR AND ANXIETY – THE BRAIN ON ALERT

Few emotions are as powerful or as universally experienced as fear. The racing heart when you hear a strange noise at night. The surge of adrenaline when a car swerves into your lane. The knot in your stomach before an important presentation. The paralysis when confronting a snake or spider. These visceral experiences of fear and anxiety are among the most intense emotional states humans can experience, and they serve crucial functions in keeping us safe from danger.

Fear is an immediate emotional response to a clear and present threat. When you encounter something dangerous, fear activates within milliseconds, preparing your body and mind to respond before conscious thought can intervene. Anxiety, in contrast, is a more diffuse state of apprehension about potential future threats. Where fear says "there is danger right now," anxiety says "something bad might happen." Both states involve similar brain systems and physiological responses, but anxiety tends to be less intense and more prolonged than fear, often persisting even when no specific threat can be identified.

From an evolutionary perspective, fear and anxiety are among our most valuable adaptations. Our ancestors who quickly detected and responded to threats like predators, poisonous plants, hostile strangers, and dangerous environments were more likely to survive and reproduce than those who were fearless or oblivious to danger. The capacity for fear has been so strongly selected throughout evolutionary history that we are born with predispositions to fear certain things like snakes, spiders, heights, and angry faces, even if we have never been harmed by them. These prepared fears likely reflect dangers that were consistently present throughout human evolution.

However, the same fear and anxiety systems that protect us from genuine dangers can malfunction, creating suffering out of proportion to actual threats. Anxiety disorders are among the most common mental health problems, affecting roughly one in five people at some point in their lives. These disorders involve fear and anxiety that are excessive in intensity or duration, triggered by stimuli that are not genuinely dangerous, or so pervasive that they interfere with daily functioning. Understanding the biological basis of fear and anxiety reveals both why these emotions are so powerful and why they sometimes spiral out of control.

This chapter explores the neuroscience of fear and anxiety in depth, examining the brain circuits that detect threats and generate defensive responses, the physiological changes that prepare the body for action, how fear memories are formed and maintained, why some fears are more easily learned than others, and what goes wrong in anxiety disorders. We will see how this biological understanding has revolutionized treatment for anxiety and trauma-related disorders, offering hope to millions of people whose lives are constrained by excessive fear.

The Amygdala: The Brain's Alarm System

As introduced in earlier chapters, the amygdala serves as the brain's primary threat detection system, constantly monitoring sensory information for signs of danger and triggering defensive responses when threats are detected. This almond-shaped structure, located deep within the temporal lobe on each side of the brain, has evolved over millions of years to perform this crucial function with remarkable speed and efficiency.

The amygdala receives inputs from all sensory systems, allowing it to evaluate visual, auditory, olfactory, tactile, and even internal bodily signals for potential threats. This convergence of sensory information makes the amygdala ideally positioned to integrate multiple sources of information and make rapid threat assessments. When you hear a loud crash, see a snake on the path, smell smoke, or feel sudden pain, this information flows to the amygdala within a few hundred milliseconds, often triggering defensive responses before you consciously identify what is happening.

The speed of amygdala responses depends partly on a shortcut pathway that bypasses detailed cortical processing. As neuroscientist Joseph LeDoux discovered through elegant experiments with rats, sensory information can reach the amygdala through two routes: a fast, crude pathway directly from the thalamus, and a slower, more detailed pathway that routes through sensory cortex. The direct thalamic pathway sacrifices accuracy for speed, conveying rough information about stimuli that allows the amygdala to begin responding in as little as twelve milliseconds in rats, far faster than conscious perception requires.

This fast pathway explains why you might jump at a garden hose that vaguely resembles a snake before consciously recognizing it is harmless. The thalamus sends crude information about a curved, elongated object to the amygdala, which triggers a fear response based on this limited information. Simultaneously, visual information routes through visual cortex, which performs detailed analysis identifying the object as a hose rather than a snake. This corrective information then reaches the amygdala through the cortical pathway, allowing the fear response to be canceled. The entire sequence from initial detection to correction might take less than a second, but the fast pathway has already protected you by initiating defensive responses before detailed identification could occur.

The amygdala is not a single uniform structure but contains multiple subdivisions with distinct functions. The lateral nucleus serves as the primary input station, receiving sensory information and detecting potentially threatening stimuli. The basal nucleus integrates information and connects to brain regions involved in memory and cognition. The central nucleus serves as the primary output station, sending signals to brain stem regions that coordinate defensive responses including freezing, fleeing, and physiological arousal. The medial nucleus is involved in social and reproductive behaviors including mate selection and parental care.

When the amygdala detects a threat, it orchestrates a coordinated defensive response involving multiple brain and body systems. The central nucleus projects to the periaqueductal gray in the mid brain, which controls defensive behaviors like freezing and flight. It projects to the lateral hypothalamus, which activates the sympathetic nervous system, producing the rapid physiological changes associated with fear including increased heart rate, elevated blood pressure, and release of adrenaline. It projects to the paraventricular nucleus of the hypothalamus, which activates the HPA axis, triggering release of stress hormones including cortisol. It projects to the locus coeruleus, which releases norepinephrine throughout the brain, increasing arousal and attention.

These widespread projections allow the amygdala to coordinate responses across multiple systems simultaneously. Within milliseconds of detecting a threat, your muscles tense for action, your heart beats faster to supply oxygen to those muscles, your breathing deepens to take in more air, your pupils dilate to improve vision, your attention narrows to focus on the threat, and stress hormones begin flooding your bloodstream to sustain this mobilization. This coordinated response represents millions of years of evolutionary refinement, optimized to promote survival when facing danger.

The amygdala's role in fear is not limited to detecting immediate threats. It also plays crucial roles in learning about danger through experience, a process called fear conditioning that allows organisms to anticipate and avoid threats based on warning signals. Understanding fear conditioning reveals how specific fears develop and why they can be so persistent and difficult to overcome.

Fear Learning: How We Acquire New Fears

Fear conditioning is one of the most extensively studied forms of learning, partly because it is simple enough to study rigorously yet powerful enough to model real-world fear acquisition. The basic paradigm involves pairing a neutral stimulus like a tone or light with an aversive stimulus like a mild electric shock. After several pairings, the previously neutral stimulus becomes a conditioned stimulus that elicits fear responses even when presented alone, without the shock.

This learning occurs rapidly and can be remarkably persistent. A single pairing of tone and shock is often sufficient to create a lasting conditioned fear response. Once established, conditioned fear can persist for months or years, demonstrating the adaptive value of quickly learning to fear warning signals of danger. An animal that touches a hot stove once and thereafter avoids similar situations has learned a valuable lesson that protects it from future burns.

The amygdala is essential for acquiring conditioned fear. Neurons in the lateral nucleus of the amygdala receive inputs carrying information about both the conditioned stimulus (the tone) and the unconditioned stimulus (the shock). When these inputs arrive close together in time, synaptic connections between them strengthen through mechanisms similar to long-term potentiation, the synaptic strengthening thought to underlie many forms of learning. After this strengthening occurs, the conditioned stimulus alone can activate amygdala neurons that trigger defensive responses.

This cellular mechanism explains several important properties of fear conditioning. The requirement for temporal contiguity between conditioned and unconditioned stimuli explains why effective conditioning requires that warning signals precede danger by only a short interval, typically a few seconds. Longer intervals produce weaker conditioning because the inputs arrive too far apart in time to produce effective synaptic strengthening. The rapid strengthening that can occur with a single pairing explains why traumatic experiences can create lasting fears. The persistence of strengthened connections explains why conditioned fears are so resistant to forgetting.

However, conditioned fear is not permanent and unchangeable. Under appropriate conditions, learned fear responses can diminish through a process called extinction. If the conditioned stimulus is repeatedly presented without the aversive outcome, fear responses gradually decline. A person who was bitten by a dog might initially fear all dogs, but repeated safe encounters with friendly dogs can extinguish this fear. Importantly, extinction does not erase the original fear memory but rather creates a new competing memory that inhibits fear expression under certain conditions.

The neural mechanisms of extinction involve brain regions beyond the amygdala, particularly the ventromedial prefrontal cortex. During extinction learning, this cortical region becomes active and appears to inhibit amygdala output, suppressing fear responses. With repeated extinction trials, connections from prefrontal cortex to amygdala strengthen, allowing cortical regions to more effectively control amygdala activity. This explains why extinction requires multiple trials and develops gradually rather than occurring as rapidly as initial fear conditioning.

The dependence of extinction on prefrontal-amygdala interactions has important implications. Because extinction creates a competing memory rather than erasing the original fear, extinguished fears can return under certain circumstances. If considerable time passes after extinction, fear responses can spontaneously recover, a phenomenon called spontaneous recovery. If the conditioned stimulus is encountered in a different context from where extinction occurred, fear may return, demonstrating that extinction is partly context-dependent. If a person experiences the aversive outcome again, even in a different context, fear to the conditioned stimulus can be rapidly reinstated.

These properties of extinction help explain why fears and phobias can be so difficult to overcome permanently. Simply avoiding a feared stimulus does not produce extinction because extinction requires exposure to the feared stimulus without adverse consequences. Even when extinction does occur through exposure, the original fear memory remains intact and can be reactivated under certain conditions. This is why exposure-based therapies for anxiety disorders typically require many sessions conducted in varied contexts to create robust extinction that generalizes across situations and resists relapse.

Prepared Fears: Not All Stimuli Are Equal

An important discovery in fear research is that not all stimuli are equally capable of becoming conditioned fear stimuli. Some classes of stimuli, particularly those that represented evolutionary threats, are much more readily associated with danger than arbitrary neutral stimuli. This phenomenon, called prepared learning or prepared conditioning, reveals how evolution has shaped learning mechanisms to facilitate rapid acquisition of fears to ancestrally relevant threats.

The psychologist Martin Seligman first proposed the preparedness theory in the 1970s to explain a puzzling observation: people develop phobias of snakes, spiders, heights, enclosed spaces, and other things that were dangerous throughout human evolution far more readily than they develop phobias of cars, guns, electrical outlets, and other modern dangers that cause far more injuries and deaths today. This selective pattern suggests that fear learning mechanisms are not blank slates that can equally associate any stimulus with danger, but rather have been shaped by evolution to favor certain associations.

Experimental research has confirmed preparedness effects. In fear conditioning studies, people more readily learn to fear pictures of snakes or spiders paired with shock than pictures of flowers or mushrooms paired with shock, even though both sets of stimuli are equally unfamiliar and the shock pairings are identical. Once established, fear conditioning to evolutionary-relevant stimuli like snakes is also more resistant to extinction than conditioning to arbitrary stimuli. Moreover, these preparedness effects occur even when stimuli are presented so briefly that participants are not consciously aware of seeing them, suggesting that prepared learning can occur outside conscious awareness.

The amygdala appears to play a special role in processing evolutionary-relevant threats. Neuroimaging studies show that the amygdala responds more strongly to pictures of snakes and spiders than to neutral objects, even in people who do not have phobias. This heightened amygdala response occurs even when snake and spider images are presented so briefly that people cannot consciously identify what they saw, demonstrating automatic preferential processing of these stimuli. People with snake or spider phobias show even greater amygdala responses, suggesting that phobias may involve exaggerated versions of normal preparedness mechanisms.

Preparedness is not limited to fear of animals and heights. Humans also show prepared learning for angry faces, which throughout evolutionary history signaled potential social threats. Studies show that angry faces are detected more quickly in visual search tasks than happy faces, and that fear conditioning to angry faces is acquired more readily and extinguished more slowly than conditioning to neutral faces. The amygdala shows preferential responses to angry and fearful faces compared to happy or neutral faces, and these responses occur even when faces are presented outside conscious awareness.

The evolutionary logic behind preparedness is compelling. Throughout the millions of years of mammalian evolution leading to humans, certain categories of stimuli consistently represented dangers: predators, venomous creatures, heights from which one could fall, unfamiliar con specifics who might be hostile, and signs of disease or contamination. Organisms with learning mechanisms that facilitated rapid acquisition of fear to these stimuli were more likely to survive and reproduce than those with unbiased learning mechanisms. Modern humans have inherited these biased learning systems even though snakes, spiders, and other evolutionary threats are far less dangerous in contemporary environments than cars, guns, and electrical hazards that have existed for only a tiny fraction of our evolutionary history.

This evolutionary perspective helps explain the distribution of phobias in the population. The most common specific phobias involve stimuli that were dangerous throughout human evolution: animals (particularly snakes, spiders, and insects), natural environments (heights, storms, water), blood-injection-injury, and enclosed spaces. In contrast, phobias of modern threats like cars and electrical outlets are virtually nonexistent despite these hazards causing far more deaths and injuries. Our brains are prepared to learn certain fears, not because those fears are objectively most useful today, but because they were most useful throughout the vast majority of our evolutionary past.

The Physiology of Fear: Preparing the Body for Danger

The subjective feeling of fear is inseparable from dramatic physiological changes throughout the body. These changes, orchestrated by the amygdala through its widespread connections, prepare the organism for defensive action. Understanding these physiological responses reveals why fear feels the way it does and why chronic anxiety can damage physical health.

The most immediate physiological response to fear involves activation of the sympathetic nervous system, the branch of the autonomic nervous system that mobilizes the body for action. Within seconds of threat detection, sympathetic nerve signals reach the heart, blood vessels, lungs, digestive system, and other organs, producing coordinated changes that optimize the body for fighting or fleeing.

The heart rate increases, sometimes dramatically, pumping more blood to deliver oxygen and nutrients to muscles that might be needed for defensive action. Blood pressure rises as blood vessels in the skin and digestive organs constrict, redirecting blood to the brain and skeletal muscles. This redistribution of blood flow explains why people often become pale when frightened, as blood drains from the skin, and why digestion slows during fear, as resources are diverted from long-term maintenance to immediate survival.

The airways in the lungs dilate, allowing more air to enter with each breath. Breathing typically becomes faster and deeper, increasing oxygen intake. These respiratory changes ensure that increased oxygen demands of activated muscles can be met. However, in the absence of physical activity, these breathing changes can lead to hyperventilation, where excessive breathing reduces carbon dioxide levels in the blood, potentially causing dizziness, tingling in the extremities, and feelings of breathlessness that can paradoxically make anxiety worse.

The pupils dilate, allowing more light to enter the eyes and potentially improving vision in the moments when detecting threats and escape routes becomes crucial. Muscle tension increases throughout the body, preparing muscles for rapid action. This tension, if sustained during chronic anxiety, contributes to muscle aches, tension headaches, and jaw pain from teeth clenching.

The adrenal medulla, the inner portion of the adrenal glands, releases adrenaline (also called epinephrine) and noradrenaline (norepinephrine) into the bloodstream. These hormones amplify and prolong sympathetic nervous system effects, ensuring that the body remains mobilized even if the initial sympathetic burst wanes. Adrenaline circulating in the blood reaches all body tissues, creating the surging, energized feeling associated with intense fear. It also affects the brain directly, enhancing memory formation for frightening events, which explains why emotional memories tend to be particularly vivid and long-lasting.

Sweating increases, particularly on the palms, soles, and underarms. This sweating serves to cool the body in anticipation of heat generated by vigorous physical activity. The moisture on palms might also improve grip, useful for climbing or wielding weapons. Changes in skin conductance caused by sweating form the basis for one component of polygraph tests, though the validity of using these responses to detect lies is questionable.

Digestive processes slow or stop, as blood and energy are diverted from long-term maintenance functions to immediate survival. This explains the loss of appetite that often accompanies fear and anxiety, and the nausea or "butterflies in the stomach" caused by reduced blood flow to the digestive system. Saliva production decreases, causing dry mouth. Bladder and bowel sphincters may relax, which in extreme cases can cause involuntary urination or defecation, a response that might reduce body weight for flight or create distractions for predators.

Beyond these immediate sympathetic responses, fear and stress activate the hypothalami-pituitary-adrenal (HPA) axis, triggering a cascade of hormonal changes that unfold over minutes to hours. The hypothalamus releases corticotropin-releasing hormone, which stimulates the pituitary gland to release adrenocorticotropic hormone, which in turn stimulates the adrenal cortex to release cortisol, the primary stress hormone.

Cortisol mobilizes energy reserves by promoting the breakdown of proteins and fats into glucose, ensuring sustained energy availability for dealing with threats. It enhances the effects of adrenaline on the cardiovascular system. It suppresses immune function and inflammatory responses, which in the short term redirects resources toward immediate survival but in the long term increases vulnerability to illness. Cortisol also affects the brain, influencing memory, attention, and mood in complex ways that depend on the amount released and the duration of elevation.

These coordinated physiological changes constitute the famous "fight-or-flight response" first described by physiologist Walter Cannon in the 1920s. Cannon recognized that the autonomic and hormonal changes accompanying fear and anger prepare the body for either defensive fighting or rapid escape, the two primary options when confronting physical threats. This mobilization system has been refined through millions of years of evolution and works beautifully for dealing with acute physical dangers.

However, as discussed in earlier chapters, modern life often presents psychological rather than physical threats work pressures, relationship conflicts, financial worries that trigger the same fight-or-flight response even though neither fighting nor fleeing resolves these problems. Moreover, these psychological stressors tend to be chronic rather than acute, keeping stress response systems activated for extended periods. This chronic activation of systems designed for brief emergency responses can cause serious health problems.

Individual Differences in Fear and Anxiety

People vary dramatically in their susceptibility to fear and anxiety. Some individuals are temperamentally anxious from early childhood, reacting to new situations with wariness and distress. Others are bold and fearless, approaching novelty with enthusiasm and showing minimal anxiety even in objectively stressful situations. These individual differences have important consequences for mental health, as people high in trait anxiety are at increased risk for developing anxiety disorders.

Behavioral inhibition, a temperamental tendency toward wariness and withdrawal when encountering unfamiliar people, places, or things, has been extensively studied by developmental psychologist Jerome Kagan and colleagues. Kagan found that about fifteen to twenty percent of infants show behavioral inhibition when confronted with novel stimuli, reacting with increased motor activity, crying, and physiological arousal. These behaviorally inhibited infants tend to remain more anxious and cautious throughout childhood compared to uninhibited peers.

Behavioral inhibition shows moderate stability across development, with inhibited infants more likely to become shy, anxious children and adolescents. However, this stability is not perfect, and many inhibited infants become less anxious over time, particularly with supportive parenting that encourages gradual exposure to new experiences without forcing or pressuring. This demonstrates that temperament sets initial tendencies but does not rigidly determine outcomes.

Neuroscience research has identified biological correlates of behavioral inhibition. Inhibited individuals tend to show greater amygdala reactivity to novel stimuli, particularly faces, suggesting their threat detection systems are more sensitive. They also show altered connectivity between the amygdala and prefrontal regulatory regions, potentially reflecting differences in the capacity to regulate anxiety responses. Some studies find altered activity in other brain regions including the anterior cingulate cortex and insula.

Genetic factors contribute substantially to individual differences in anxiety, with heritability estimates typically ranging from thirty to fifty percent for various measures of trait anxiety and specific anxiety disorders. This means that genetic differences among individuals account for roughly thirty to fifty percent of population variation in anxiety, with environmental factors accounting for the remainder. Twin studies, which compare identical twins who share all their genes with fraternal twins who share only half, consistently show greater similarity in anxiety measures for identical twins, confirming genetic influence.

However, no single gene determines anxiety. Instead, many genes, each with small individual effects, combine with environmental factors to influence risk. Some of the genes implicated in anxiety affect serotonin, GABA, and other neurotransmitter systems involved in fear and anxiety. Others affect stress hormone regulation, neural plasticity, or other processes relevant to anxiety. The complexity of genetic influences helps explain why anxiety runs in families but does not follow simple inheritance patterns.

Environmental factors also play crucial roles in shaping anxiety. Stressful life events, particularly during childhood, increase risk for anxiety disorders. Children who experience abuse, neglect, parental loss, or other adverse events show elevated rates of anxiety disorders in adulthood. Animal research demonstrates that early-life stress produces lasting changes in stress response systems, including altered HPA axis function and increased amygdala reactivity, providing biological mechanisms linking early adversity to later anxiety.

Parenting styles also influence anxiety development. Overprotective parenting that shields children from all stress and discourages independent coping can paradoxically increase anxiety by preventing children from learning that they can handle challenges. Parental modeling of anxious responses teaches children to interpret situations as threatening. Conversely, warm, supportive parenting that encourages gradual mastery of age-appropriate challenges promotes resilience and reduces anxiety risk.

Understanding the interplay of genetic and environmental factors in anxiety helps explain why some people develop anxiety disorders while others exposed to similar stressors remain healthy. Genetic factors create differential susceptibility to environmental influences, with some individuals more vulnerable to stress but also potentially more responsive to positive interventions. This gene-environment interaction perspective suggests that anxiety prevention and treatment should be tailored to individual characteristics rather than applying one-size-fits-all approaches.

When Fear Becomes Pathological: Anxiety Disorders

While fear and anxiety serve adaptive functions, they can become excessive or inappropriate, creating the various anxiety disorders that cause immense suffering and impairment. These disorders are among the most common mental health problems, affecting hundreds of millions of people worldwide, yet they often go untreated because sufferers are ashamed to seek help or do not recognize that effective treatments exist.

Specific phobias involve intense, irrational fears of particular objects or situations such as animals, heights, flying, enclosed spaces, blood, or medical procedures. The fear is disproportionate to actual danger, persists despite recognition that it is excessive, and leads to avoidance that can significantly interfere with daily life. A person with a snake phobia might refuse to go hiking or even work in their garden for fear of encountering snakes. Someone with a flying phobia might turn down job promotions requiring travel or miss important family events requiring air travel.

Phobias can develop through several pathways. Some arise from direct traumatic experiences, such as developing a dog phobia after being bitten. Others are acquired through observation, as when a child develops a fear of spiders after watching a parent react with terror to spiders. Still others seem to arise without any clear learning experience, possibly reflecting prepared fears that require little learning to become activated. Regardless of origin, phobias persist through avoidance, which prevents extinction learning that would occur if the person repeatedly encountered the feared stimulus without adverse consequences.

Social anxiety disorder, previously called social phobia, involves intense fear of social situations where one might be scrutinized or evaluated by others. People with social anxiety disorder fear embarrassing themselves, being judged negatively, or showing visible signs of anxiety such as blushing, trembling, or sweating. These fears lead to avoidance of social situations or endurance of them with intense distress. Social anxiety disorder typically begins in adolescence and can severely restrict educational and career opportunities, romantic relationships, and quality of life.

The biology of social anxiety disorder involves heightened threat detection in response to social-evaluative situations. The amygdala shows increased reactivity to faces, particularly critical or angry faces. Regions involved in self-referential processing and error monitoring, including the medial prefrontal cortex and anterior cingulate cortex, show altered activity, potentially reflecting excessive self-focused attention and monitoring for mistakes. Neurotransmitter systems implicated in social anxiety include serotonin, dopamine, and neuropeptides including oxytocin that regulate social behavior.

Panic disorder is characterized by recurrent unexpected panic attacks, intense surges of fear accompanied by physical symptoms including rapid heartbeat, sweating, trembling, shortness of breath, chest pain, nausea, dizziness, and feelings of losing control or dying. Panic attacks reach peak intensity within minutes and typically subside within twenty to thirty minutes, though they may leave the person exhausted and shaken. Beyond the attacks themselves, panic disorder involves persistent worry about having additional attacks and behavioral changes intended to prevent attacks.

The neurobiology of panic disorder involves abnormalities in brain regions that monitor physiological states and coordinate defensive responses. The periaqueductal gray, a midbrain structure that orchestrates defensive behaviors including freezing and escape, may be hypersensitive. The insular cortex, which integrates information about bodily states, may misinterpret normal physiological fluctuations as signs of danger, triggering panic. Respiratory control systems may be dysregulated, with heightened sensitivity to carbon dioxide potentially contributing to hyperventilation and breathlessness during panic attacks.

An influential theory proposes that panic attacks result from catastrophic misinterpretation of bodily sensations. According to this cognitive model, people prone to panic attacks interpret normal bodily fluctuations like increased heart rate or slight breathlessness as signs of imminent catastrophe such as a heart attack, suffocation, or loss of control. These catastrophic thoughts trigger genuine fear, which produces the very physiological symptoms that are being misinterpreted, creating a vicious cycle. The person notices their heart beating faster, interprets this as a sign of a heart attack, becomes terrified, which makes their heart beat even faster, confirming their fears and escalating panic.

This cognitive model explains why panic attacks often seem to come "out of the blue" with no obvious external trigger. Internal bodily sensations serve as the trigger, but because these sensations are not typically consciously noticed, the resulting panic seems unprovoked. The model also explains why panic disorder often develops after an initial panic attack that occurred for some reason such as caffeine over consumption, sleep deprivation, or illness. The terrifying experience of the first attack creates hypervigilance for bodily sensations, setting the stage for subsequent panic attacks triggered by catastrophic misinterpretation of normal physiological fluctuations.

Generalized anxiety disorder involves excessive worry about multiple domains of life that is difficult to control and causes significant distress or impairment. Unlike specific phobias with clear feared objects or panic disorder with discrete attacks, generalized anxiety disorder involves chronic, diffuse worry about work, health, finances, relationships, and everyday responsibilities. The worry is often accompanied by physical symptoms including restlessness, fatigue, difficulty concentrating, irritability, muscle tension, and sleep disturbance.

The neurobiology of generalized anxiety disorder appears to involve dysfunction in neural circuits that regulate worry and threat monitoring. The amygdala shows elevated baseline activity and heightened responses to potential threats. The prefrontal cortex, particularly regions involved in cognitive control and emotion regulation, shows altered activity and may struggle to inhibit anxious thoughts once they begin. Some research suggests reduced GABA inhibitory function throughout the brain, potentially explaining the diffuse nature of symptoms.

Obsessive-compulsive disorder, commonly abbreviated OCD, involves recurrent intrusive thoughts, images, or urges (obsessions) that cause anxiety, and repetitive behaviors or mental acts (compulsions) performed to reduce the anxiety caused by obsessions. Common obsessions include fears of contamination, fears of causing harm through negligence, need for symmetry or exactness, and forbidden sexual or religious thoughts. Common compulsions include excessive hand washing, checking locks and appliances repeatedly, counting, arranging objects in precise order, and mental rituals like praying or repeating phrases.

OCD was once classified as an anxiety disorder but is now recognized as a distinct condition, though anxiety certainly plays a central role. The neurobiology of OCD involves dysfunction in circuits connecting the orbitofrontal cortex, anterior cingulate cortex, striatum, and thalamus. These circuits normally signal when a goal has been achieved and allow termination of goal-directed behavior. In OCD, these circuits appear to malfunction, failing to generate the feeling that a goal has been adequately accomplished. This creates a persistent sense that something is not right or not complete, driving repetitive checking and ritualistic behaviors that temporarily reduce anxiety but never fully resolve the underlying feeling of incompleteness.

Neuroimaging studies consistently show hyperactivity in the orbitofrontal cortex and anterior cingulate cortex in OCD patients, with activity levels often correlating with symptom severity. Successful treatment with either medication or cognitive-behavioral therapy reduces activity in these regions toward normal levels, providing evidence that the observed hyperactivity is related to symptoms rather than simply being an incidental finding. Neurotransmitter systems implicated in OCD include serotonin, dopamine, and glutamate, with medications affecting these systems often providing symptom relief.

Post-traumatic stress disorder, discussed briefly in the previous chapter, warrants further attention in the context of fear and anxiety. PTSD develops in some people following exposure to traumatic events and involves intrusive re-experiencing of the trauma through flashbacks and nightmares, avoidance of trauma reminders, negative alterations in cognition and mood, and heightened arousal and reactivity. PTSD represents a failure of normal fear memory processing, with trauma memories remaining hyperaccessible and continuing to trigger intense fear responses long after the original danger has passed.

The biology of PTSD involves multiple abnormalities in fear processing systems. The amygdala shows heightened reactivity to trauma reminders and potentially to neutral stimuli, reflecting overgeneralized fear responses. The hippocampus, crucial for contextual memory, often shows reduced volume and altered function, potentially contributing to difficulty distinguishing safe from dangerous contexts. The ventromedial prefrontal cortex, essential for fear extinction, shows reduced activity, potentially explaining why learned fear responses fail to extinguish naturally with time and safe re-exposure to trauma-related cues.

Trauma memories in PTSD have unusual qualities that distinguish them from normal memories. They often feel more like current perceptions than remembered events, with sensory details experienced with vivid immediacy. They intrude into consciousness involuntarily rather than being deliberately recalled. They may lack the normal narrative structure of memories, appearing instead as disconnected sensory fragments. These qualities suggest that trauma memories are processed and stored differently from normal memories, possibly because intense stress disrupts normal memory consolidation processes.

The hormonal environment during trauma may contribute to these memory abnormalities. Extremely high levels of stress hormones including cortisol and norepinephrine during trauma enhance emotional aspects of memory while potentially impairing hippocampal-dependent contextual processing. This creates vivid sensory and emotional memories that lack adequate contextual information about when and where the trauma occurred and that it is now over. The resulting memory feels timeless and current rather than being properly marked as belonging to the past.

The Neuroscience of Treatment: How Therapy Changes the Brain

Understanding the neurobiology of fear and anxiety has revolutionized treatment, leading to therapies specifically designed to target the brain mechanisms maintaining these conditions. The most effective psychological treatments for anxiety disorders involve exposure to feared stimuli in safe contexts, taking advantage of extinction learning mechanisms to reduce fear responses.

Exposure therapy, used to treat phobias, social anxiety, panic disorder, and PTSD, involves systematic confrontation with feared stimuli in a gradual, controlled manner. For a person with a dog phobia, treatment might begin with looking at pictures of dogs, progress to watching videos of dogs, then to observing a calm dog from a distance, then to being in the same room with a leashed dog, and eventually to touching and interacting with dogs. Each exposure continues until anxiety naturally decreases, allowing extinction learning to occur.

The neuroscience of exposure therapy reveals how this treatment works at the brain level. Repeated exposure to feared stimuli without adverse consequences activates extinction learning processes in the ventromedial prefrontal cortex. This region sends inhibitory signals to the amygdala, reducing fear responses. With repeated successful exposures, connections from prefrontal cortex to amygdala strengthen, making it easier to inhibit fear in future encounters with the feared stimulus.

Brain imaging studies have documented these neural changes. Before exposure therapy, patients with phobias or PTSD show exaggerated amygdala responses to their specific fear triggers and reduced prefrontal cortex activity. After successful exposure therapy, amygdala responses decrease and prefrontal activity increases, particularly in regions involved in extinction. These neural changes correlate with symptom reduction, providing direct evidence that therapy works by altering brain function.

Understanding extinction as new learning rather than erasure of original fear memories has important implications for treatment. Because extinction is context-dependent, exposure therapy should ideally occur in multiple contexts to promote generalization. Because extinguished fears can return through spontaneous recovery, reinstatement, or context change, patients need strategies for managing fear if it returns rather than interpreting return of symptoms as treatment failure. Booster sessions of exposure therapy can strengthen extinction and prevent relapse.

Cognitive-behavioral therapy combines exposure with cognitive restructuring, teaching patients to identify and challenge anxious thoughts. For panic disorder, therapy helps patients recognize that the bodily sensations they fear are not dangerous, breaking the cycle of catastrophic misinterpretation. For social anxiety, therapy challenges beliefs about the likelihood and consequences of negative evaluation. For generalized anxiety disorder, therapy addresses excessive worry and teaches skills for tolerating uncertainty.

The neurobiological effects of cognitive therapy likely involve prefrontal regions that regulate emotional responses through reappraisal and cognitive control. Learning to reinterpret bodily sensations as harmless rather than dangerous, or to view social situations as opportunities rather than threats, requires prefrontal regions to override automatic amygdala responses with more adaptive interpretations. Brain imaging studies show that cognitive therapy increases activity in dorsolateral and ventromedial prefrontal cortex while reducing amygdala reactivity, consistent with enhanced cognitive regulation of emotion.

Medications for anxiety disorders work through different mechanisms than psychotherapy but ultimately affect some of the same brain systems. Selective serotonin re uptake inhibitors, widely used for anxiety disorders, increase serotonin availability throughout the brain. Over weeks of treatment, this appears to promote neural plasticity and improve prefrontal regulation of amygdala activity, mechanisms partially overlapping with those of psychotherapy. This may explain why combining medication with cognitive-behavioral therapy is often more effective than either treatment alone.

Benzodiazepines, medications that enhance GABA inhibitory neurotransmission, provide rapid anxiety relief but have significant drawbacks including sedation, cognitive impairment, and risk of dependence. They work by increasing inhibition throughout the brain, essentially turning down neural activity globally rather than specifically targeting anxiety circuits. While useful for short-term management of severe anxiety, they are generally not recommended for long-term treatment because tolerance develops, requiring increasing doses for the same effect, and withdrawal can be dangerous.

Beta-blockers, medications that block adrenaline receptors, can reduce the physical symptoms of anxiety including rapid heartbeat and trembling. They are sometimes used for performance anxiety in situations like public speaking or musical performances. By reducing physical symptoms, beta-blockers may prevent the feedback loop where awareness of physical arousal increases anxiety, which further increases arousal. However, they do not address the psychological aspects of anxiety and are not effective for most anxiety disorders.

Novel treatments based on neuroscience insights continue to emerge. D-cycloserine, a medication that enhances learning by facilitating NMDA receptor function, can strengthen extinction learning when given shortly before or after exposure therapy sessions. This demonstrates that pharmacological enhancement of neural plasticity can augment psychotherapy effects. Research on this approach continues, with the goal of identifying optimal dosing and timing to maximize benefits.

Re consolidation-based therapies represent another innovative approach. Research shows that when a memory is recalled, it enters a temporary labile state during which it can be modified before being re consolidated into long-term storage. Interventions during this re consolidation window might allow permanent modification of fear memories. Some studies suggest that recalling a fear memory and then engaging in extinction training during the re consolidation window produces more lasting fear reduction than standard extinction. This approach remains experimental but offers exciting possibilities for more effective trauma treatment.

Virtual reality exposure therapy uses immersive computer-generated environments to conduct exposure therapy, particularly useful for fears that are difficult to confront in real life such as flying, heights, or combat situations in PTSD treatment. Virtual reality allows graduated exposure with precise therapist control, and research suggests it can be as effective as real-world exposure for many conditions. Brain imaging studies show that virtual reality exposure activates similar brain regions as real exposure, supporting its validity as a treatment tool.

Stress, Fear, and Physical Health

The connection between psychological stress and physical illness has long been recognized, but understanding the biological mechanisms linking mind and body has revealed just how profound this connection is. Chronic activation of fear and stress systems damages health through multiple pathways, affecting cardiovascular function, immune responses, metabolic processes, and even brain structure.

The cardiovascular effects of chronic stress are particularly well-documented. Repeated activation of the sympathetic nervous system and sustained elevation of stress hormones keep blood pressure elevated, heart rate increased, and blood vessels constricted. Over time, these chronic changes damage blood vessel walls, promote atherosclerosis (hardening of the arteries), and increase risk of heart attack and stroke. People experiencing chronic stress show accelerated progression of cardiovascular disease compared to those with similar risk factors but lower stress levels.

The immune system, which defends against infections and cancer, is profoundly affected by stress. Acute stress temporarily enhances certain immune functions, mobilizing defenses in preparation for potential injuries. However, chronic stress suppresses immune function through multiple mechanisms including elevated cortisol, which has anti-inflammatory and immunosuppressive effects. People under chronic stress show increased susceptibility to infections, slower wound healing, and reduced responses to vaccines.

Pioneering research by psychologist Janice Kiecolt-Glaser and immunologist Ronald Glaser demonstrated immune suppression in medical students during examination periods, in people caring for family members with dementia, and in couples experiencing marital conflict. These studies established that psychological stress produces measurable changes in immune function that have real health consequences. Chronic stress is associated with increased inflammation, which contributes to numerous diseases including cardiovascular disease, diabetes, arthritis, and depression.

The metabolic effects of chronic stress include disrupted glucose regulation and increased abdominal fat accumulation. Cortisol promotes glucose production and reduces insulin sensitivity, which over time can contribute to development of type 2 diabetes. Stress also affects eating behavior, with many people eating more high-calorie comfort foods when stressed, and alters how the body stores fat, favoring accumulation of visceral fat around internal organs. This visceral fat is particularly harmful, producing inflammatory molecules and increasing cardiovascular and metabolic disease risk.

In the brain, chronic stress produces structural changes that impair cognitive function and increase vulnerability to mental illness. The hippocampus is particularly vulnerable to stress, with chronic stress causing atrophy and reduced neurogenesis (birth of new neurons) in this region. These changes impair memory and may contribute to the memory problems often reported by people with chronic stress and depression. The prefrontal cortex also shows stress-induced atrophy and reduced connectivity, potentially impairing executive functions including decision-making, planning, and emotion regulation.

Conversely, the amygdala shows increased volume and heightened reactivity with chronic stress, creating a brain increasingly biased toward threat detection and defensive responding. This combination of enhanced amygdala function, reduced hippocampal and prefrontal function, creates a brain less able to regulate emotions, less able to distinguish safe from dangerous contexts, and more prone to anxiety and depression.

These stress-induced brain changes are not necessarily permanent. Research on stress management interventions including exercise, meditation, and stress reduction therapy shows that these practices can reverse some stress-related brain changes, promoting hippocampal neurogenesis, increasing prefrontal cortex thickness, and reducing amygdala reactivity. This neuroplasticity provides biological mechanisms explaining how behavioral interventions protect mental and physical health.

Resilience: Why Some People Cope Better with Stress

A puzzling observation in stress research is that people vary tremendously in how they respond to stressful experiences. Some individuals develop anxiety disorders, depression, or PTSD following trauma or chronic stress, while others exposed to similar or even worse stressors remain psychologically healthy. Understanding resilience, the ability to adapt successfully despite adversity, has become a major focus of research with important implications for prevention and treatment.

Resilience is not simply the absence of vulnerability but involves active processes that promote positive adaptation. Resilient individuals tend to have certain characteristics including flexible thinking, strong social support, sense of purpose and meaning, active coping strategies, and positive emotions even during difficult times. These psychological characteristics have biological correlates that help explain resilience at the neural level.

One important neurobiological factor in resilience is neuropeptide Y, a neurotransmitter that counteracts stress responses. People with higher levels of neuropeptide Y show better stress resilience, less anxiety, and quicker recovery from stress. Special operations soldiers, who undergo extremely stressful training and missions but are selected for stress resilience, show higher neuropeptide Y levels than average. Animal research shows that enhancing neuropeptide Y reduces anxiety-like behaviors and promotes active coping with stress.

The balance between stress hormone systems and their regulation also contributes to resilience. While chronic stress typically involves sustained cortisol elevation, resilient individuals often show more dynamic cortisol responses with rapid increases during acute stress followed by efficient return to baseline. This pattern suggests effective stress system activation and termination rather than the blunted or dysregulated patterns seen in some stress-related disorders.

Brain structure and function also differ in resilient individuals. Some research suggests that resilient people have larger or more active prefrontal regions involved in emotion regulation and cognitive reappraisal. They may also show stronger connectivity between prefrontal regions and the amygdala, providing more effective top-down regulation of threat responses. These brain differences might be partially genetic but can also be enhanced through experience and training.

Early life experiences profoundly influence stress resilience. Secure attachment relationships in childhood, where caregivers provide reliable comfort and support, promote development of effective stress regulation systems. Children who experience responsive care giving during stress learn that stress is manageable and that help is available, creating working models of the world as ultimately safe and controllable. These early experiences shape stress biology, with securely attached children showing more adaptive cortisol responses and better emotion regulation throughout life.

Conversely, adverse childhood experiences including abuse, neglect, parental loss, and household dysfunction increase vulnerability to stress-related disorders in adulthood. These experiences can alter development of stress response systems, creating heightened reactivity and impaired regulation that persists into adulthood. However, even among people with significant early adversity, many prove resilient, often supported by stable relationships with other adults, engagement in meaningful activities, and development of skills and competencies that build self-efficacy.

Social support emerges as one of the most powerful protectors against stress. People with strong social connections show reduced stress hormone responses, better immune function, lower rates of anxiety and depression, and even lower mortality rates compared to socially isolated individuals. Social support appears to buffer stress through multiple mechanisms including providing practical assistance with problems, offering emotional comfort during distress, and enhancing sense of meaning and belonging.

The biological mechanisms underlying social support effects include the neuropeptide oxytocin, often called the bonding hormone. Oxytocin is released during positive social interactions including physical touch, conversation with trusted others, and caregiving. It reduces stress hormone release, decreases amygdala reactivity, and promotes feelings of calm and connection. Oxytocin may be one mechanism through which social support confers its health benefits.

Understanding resilience has shifted prevention efforts from simply trying to reduce risk factors and stress exposure toward actively building protective factors and coping skills. Resilience training programs teach stress management techniques including cognitive reappraisal, relaxation skills, physical exercise, maintenance of social connections, and cultivation of purpose and meaning. While not everyone exposed to these programs develops equal resilience, many people benefit from learning and practicing these skills, demonstrating that resilience can be enhanced through intervention.

Fear and Anxiety Across the Lifespan

The expression and experience of fear and anxiety change dramatically across development, reflecting maturation of brain systems, cognitive abilities, and life experiences. Understanding these developmental changes provides insight into when different fears are normal versus pathological and how to most effectively address anxiety at different ages.

Infants show fear responses from birth, but the objects of fear change as cognitive abilities develop. Very young infants show distress to loud noises, sudden movements, and loss of physical support, fears that require minimal cognitive processing and likely reflect innate defensive responses. Around seven to nine months, stranger anxiety emerges as infants develop the ability to distinguish familiar from unfamiliar people and form specific attachments. Separation anxiety peaks around twelve to eighteen months as infants develop object permanence, the understanding that things continue to exist even when out of sight, making them aware that caregivers who leave might not return.

During the preschool years, fears often involve imaginary creatures, darkness, animals, and natural phenomena like thunder. These fears reflect developing imagination and cognitive abilities that allow children to anticipate potential dangers, but they outpace understanding of what is truly dangerous versus merely unfamiliar or unusual. Most childhood fears are transient, diminishing as children gain experience and understanding.

School-age children develop more realistic fears reflecting better understanding of genuine dangers. Fears of injury, illness, death, and school performance become more common. Social fears also increase as peer relationships become more important and children become more aware of social evaluation. Some children develop anxiety about separating from parents when going to school or other activities, called separation anxiety disorder when it is excessive and impairing.

Adolescence brings heightened emotional reactivity and increased vulnerability to anxiety disorders, particularly social anxiety disorder. The brain changes occurring during adolescence, including increased amygdala reactivity combined with still-developing prefrontal regulatory capacity, contribute to emotional volatility. Adolescents also face increasing social pressures, academic demands, and identity challenges that can trigger or exacerbate anxiety. Many adult anxiety disorders have their onset during adolescence, making this a critical period for early identification and intervention.

Young adulthood through middle age is characterized by numerous life stressors including educational pursuits, career development, romantic relationships, marriage, parenting, and financial pressures. These stressors can trigger anxiety disorders in vulnerable individuals, particularly generalized anxiety disorder. However, this period also brings continued development of coping skills and emotion regulation, and many people show improved anxiety management as they gain life experience.

Older adulthood presents unique challenges for anxiety. While anxiety disorders are less prevalent in older adults than in younger adults, they remain common and often go unrecognized and untreated. Older adults face stressors including health problems, loss of independence, bereavement, and confronting mortality. Anxiety in older adults often presents atypically, with more focus on physical symptoms than worry, potentially leading to misattribution of anxiety to medical conditions.

Brain changes in aging affect anxiety in complex ways. Some changes might reduce anxiety, including reduced amygdala reactivity to negative stimuli and improved emotion regulation in healthy older adults. However, age-related changes in prefrontal cortex function and neurotransmitter systems might increase vulnerability to certain anxiety symptoms. Cognitive decline can impair the ability to use cognitive coping strategies that were previously effective. Medical conditions and medications common in older adults can also affect anxiety.

Treatment of anxiety across the lifespan requires developmental sensitivity. For children, treatment often involves parents, with parent training in managing child anxiety being highly effective. Exposure therapy for children requires age-appropriate methods and careful attention to the child's developmental level. For adolescents, treatment must navigate issues of autonomy and peer influence while still involving families. For older adults, treatment must address medical comorbidities, accommodate cognitive limitations, and recognize the realities of age-related losses rather than simply trying to eliminate all anxiety.

The Future of Fear and Anxiety Research

Research on the neurobiology of fear and anxiety continues to advance rapidly, with new discoveries promising improved understanding and treatment. Optogenetics, a technology that allows researchers to activate or inhibit specific types of neurons using light, has revolutionized animal research on fear circuits. Scientists can now test causal relationships between activity in specific neural populations and fear behavior, providing unprecedented precision in understanding how circuits function.

These studies have revealed that even within the amygdala, distinct populations of neurons encode different aspects of fear. Some neurons specifically encode fear to particular stimuli. Others encode safety signals that inhibit fear. Still others encode uncertainty or ambiguity. Understanding this complexity at the cellular level may eventually enable more targeted treatments that affect specific aspects of fear processing while leaving others intact.

Advances in human brain imaging, including higher resolution scanners and improved analysis methods, are revealing finer details of fear circuit function in humans. Researchers can now examine connectivity between specific sub regions within the amygdala and prefrontal cortex, activity in small brain stem nuclei that coordinate defensive responses, and real-time dynamics of circuit interactions during fear learning and extinction.

Genetic and molecular studies continue to identify genes and cellular mechanisms involved in fear and anxiety. Understanding genetic contributions may eventually enable personalized medicine approaches where treatment selection is based on individual genetic profiles. Identifying molecular mechanisms of fear learning and extinction may reveal new drug targets for enhancing therapeutic extinction or preventing pathological fear formation after trauma.

The integration of computational approaches with neuroscience is enabling development of formal models of how fear circuits perform computations. These models can make testable predictions about neural activity and behavior, driving more hypothesis-driven research. Computational psychiatry applies these approaches to understand how circuit dysfunctions produce symptoms of anxiety disorders, potentially leading to better diagnostic classification and treatment selection.

Translation of basic neuroscience findings into clinical applications remains a major goal. While our understanding of fear neurobiology has advanced dramatically, this knowledge has not yet fully transformed clinical practice. Future research must bridge the gap between laboratory discoveries and clinical implementation, testing whether interventions based on neuroscience principles actually improve outcomes for patients with anxiety disorders.

Conclusion: Living with Fear in a Modern World

Fear and anxiety are fundamental aspects of human biology, systems refined over millions of years to protect our ancestors from the dangers of their environments. These same systems continue to serve important protective functions today, alerting us to genuine dangers and motivating appropriate caution. However, the mismatch between ancestral environments that shaped our fear systems and the modern world in which we live creates challenges.

Many of the dangers that shaped our evolution predators, hostile strangers, poisonous animals are rarely encountered in modern life. Meanwhile, many significant modern dangers automobiles, firearms, unhealthy diets, sedentary lifestyles fail to trigger strong fear responses because they were not present throughout our evolutionary history. This creates the paradox that we are prepared to fear snakes and spiders that rarely harm us while feeling comfortable with cars and junk food that cause enormous numbers of deaths.

Moreover, modern life presents chronic psychological stressors work pressures, financial insecurity, information overload, social comparison through social media that activate stress systems designed for acute physical dangers. The resulting chronic stress contributes to epidemic levels of anxiety disorders and stress-related physical illness. Learning to manage fear and anxiety in ways that honor their adaptive functions while preventing their pathological expression is one of the great challenges of modern life.

Understanding the biology of fear and anxiety provides tools for meeting this challenge. It explains why anxiety disorders are not moral failings or signs of weakness but rather dysfunctions in brain systems operating outside conscious control. It validates the experiences of people suffering from anxiety and reduces the stigma that prevents many from seeking help. It guides development of effective treatments based on knowledge of neural mechanisms. It suggests prevention strategies including stress management, exercise, social connection, and resilience training.

7

STRESS, ANXIETY, AND CORTISOL

Stress is one of the most universally experienced yet poorly understood aspects of human life. Everyone knows what it feels like to be stressed the tight muscles, the racing thoughts, the sense of being overwhelmed, the fatigue that comes from constantly feeling under pressure. Yet stress is far more than just an unpleasant subjective experience. It is a complex biological response involving coordinated changes across the brain, nervous system, endocrine system, and immune system. These changes, which evolved to help organisms survive acute physical threats, can become profoundly harmful when activated chronically in response to the psychological pressures of modern life.

The word stress itself has become so overused that it risks losing meaning. People describe themselves as stressed about everything from major life crises to minor inconveniences, from genuine threats to imagined dangers. This linguistic imprecision obscures important distinctions between different types of stress and their effects. A job interview, a car accident, chronic financial insecurity, caring for an ill family member, and reading disturbing news all trigger stress responses, but they differ dramatically in intensity, duration, and controllability. Understanding these differences is crucial for understanding when stress is adaptive versus harmful.

The physiologist Hans Selye, who pioneered scientific study of stress in the mid-twentieth century, defined stress as the nonspecific response of the body to any demand placed upon it. Selye recognized that widely different stressors heat, cold, injury, infection, emotional distress all produce similar patterns of physiological response involving activation of the sympathetic nervous system and the hypothalami-pituitary-adrenal axis. This common response pattern, which Selye called the general adaptation syndrome, represented the body's attempt to maintain stability in the face of challenge.

However, Selye's definition of stress as nonspecific response proved too broad. Modern stress research distinguishes between stressors, the events or circumstances that create demands, and stress responses, the physiological and psychological reactions to those demands. The relationship between stressors and responses is not simple or direct. The same stressor can produce vastly different responses in different people or even in the same person at different times, depending on how the stressors is appraised, what resources are available for coping, and what prior experiences have shaped stress response systems.

This chapter explores the biology of stress in depth, examining the brain systems that evaluate threats and initiate stress responses, the cascade of neural and hormonal changes that mobilize the body for action, how chronic stress damages both physical and mental health, why people differ in their vulnerability to stress, and what interventions can protect against stress-related illness. We will see that stress is not simply an inevitable consequence of modern life but rather a biological response that can be understood, managed, and even harnessed to promote growth and resilience.

The Stress Response: From Perception to Physiology

Stress responses begin not with objective characteristics of situations but with how those situations are perceived and evaluated. The process of stress appraisal, first described by psychologist Richard Lazarus, involves two crucial judgments. Primary appraisal evaluates whether a situation is irrelevant, benign-positive, or stressful. If a situation is appraised as stressful, further evaluation determines whether it represents harm or loss that has already occurred, threat of future harm, or challenge that can be overcome through effort. Secondary appraisal evaluates what resources are available for coping with the stressors and whether those resources are adequate.

These appraisals largely determine whether and how intensely stress responses are activated. A job interview might be appraised as a terrible threat by someone who desperately needs work and doubts their qualifications, triggering intense stress. The same interview might be appraised as an interesting challenge by someone who has other options and feels confident, producing minimal stress and perhaps even excitement. A mountain hike might be stressful for someone afraid of heights but relaxing for someone who enjoys outdoor activities. The objective situations are identical, but their psychological meaning differs dramatically.

The brain regions involved in stress appraisal include the prefrontal cortex, which evaluates situations and their potential consequences; the anterior cingulate cortex, which detects conflicts and signals the need for increased control; and the amygdala, which rapidly detects potential threats and assigns emotional significance to stimuli. These regions work together to determine whether a situation requires mobilization of stress response systems.

When a situation is appraised as stressful and requiring action, the hypothalamus initiates coordinated stress responses through two major pathways. The first pathway, already discussed in earlier chapters, involves the sympathetic nervous system and produces rapid physiological changes within seconds. The hypothalamus sends signals to the brain stem, which activates sympathetic preganglionic neurons in the spinal cord. These neurons trigger a cascade of responses including increased heart rate and blood pressure, dilated pupils, inhibited digestion, released glucose from energy stores, and activation of sweat glands.

The sympathetic nervous system also triggers the adrenal medulla, located in the center of the adrenal glands atop the kidneys, to release the hormones epinephrine (adrenaline) and norepinephrine (noradrenaline) into the bloodstream. These hormones circulate throughout the body, amplifying and prolonging sympathetic effects. Epinephrine in particular creates the surge of energy and alertness characteristic of acute stress, sometimes called an adrenaline rush. This rapid sympathetic response prepares the body for immediate action, whether fighting, fleeing, or freezing in response to threat.

The second pathway involves the hypothalami-pituitary-adrenal axis, commonly abbreviated HPA axis, which produces slower but more sustained stress responses. The paraventricular nucleus of the hypothalamus releases corticotropin-releasing hormone (CRH) into specialized blood vessels that carry it to the anterior pituitary gland. CRH stimulates pituitary cells to release adrenocorticotropic hormone (ACTH) into general circulation. ACTH travels through the bloodstream to the adrenal cortex, the outer portion of the adrenal glands, where it stimulates synthesis and release of glucocorticoid hormones, primarily cortisol in humans.

This cascade of hormones from hypothalamus to pituitary to adrenal cortex takes several minutes to reach peak levels, much slower than the sympathetic response that begins within seconds. However, once cortisol is released, it persists in the bloodstream for hours, providing sustained mobilization of energy and other stress-related changes. This slower, sustained response complements the rapid sympathetic response, ensuring that the body remains prepared to deal with prolonged challenges.

Cortisol: The Stress Hormone

Cortisol is often called the stress hormone because its levels rise markedly during stress, and it mediates many of the physiological effects of chronic stress. However, this designation is somewhat misleading because cortisol serves many important functions beyond stress responses and is essential for normal health. Understanding both the adaptive functions of cortisol and how excessive or prolonged elevation becomes harmful is crucial for understanding stress effects on health.

Under normal circumstances, cortisol follows a pronounced daily rhythm called the diurnal rhythm or circadian rhythm. Cortisol levels are highest in the early morning, typically peaking within thirty to forty-five minutes after waking, a phenomenon called the cortisol awakening response. This morning surge of cortisol helps you transition from sleep to wakefulness, mobilizing energy and promoting alertness. Throughout the day, cortisol levels gradually decline, reaching their lowest point in the evening and during the first half of the night. This pattern supports the natural sleep-wake cycle, with high morning cortisol promoting wakefulness and low evening cortisol allowing sleep.

Beyond this daily rhythm, cortisol helps regulate numerous physiological processes. It promotes gluconeogenesis, the synthesis of glucose from proteins and fats, ensuring adequate blood sugar levels especially during fasting periods like overnight sleep. It regulates immune function, generally suppressing excessive inflammatory responses that could damage tissues. It affects bone formation, with chronically high levels reducing bone density. It influences cardiovascular function, helping maintain blood pressure and vascular tone. It affects brain function, modulating memory, attention, and mood.

During acute stress, the sharp rise in cortisol serves adaptive functions. Cortisol mobilizes energy by promoting breakdown of proteins and fats into glucose, providing fuel for dealing with challenges. It enhances the effects of epinephrine on cardiovascular function, helping maintain blood pressure and cardiac output. It suppresses non-essential functions including digestion, growth, and reproduction, redirecting resources toward immediate survival. It affects memory, typically enhancing memory for stressful events while impairing retrieval of previously learned information, a pattern that focuses attention on current threats while preventing distraction by irrelevant memories.

Cortisol also affects the brain regions involved in stress responses, creating feedback loops that normally regulate and limit stress responses. Cortisol binds to glucocorticoid receptors in the hippocampus, hypothalamus, and pituitary gland. In the hypothalamus and pituitary, cortisol inhibits further release of CRH and ACTH, creating negative feedback that limits cortisol secretion once adequate levels have been reached. This feedback mechanism prevents excessive cortisol release and helps terminate stress responses once threats have passed.

However, chronic stress disrupts this regulatory system in ways that can be harmful. With repeated or prolonged stress, cortisol levels remain elevated for extended periods, exposing tissues throughout the body to high glucocorticoid levels that these tissues are not designed to sustain. Moreover, chronic stress can alter the sensitivity of glucocorticoid receptors, sometimes reducing their responsiveness in a process called glucocorticoid resistance. This reduced sensitivity impairs negative feedback, allowing cortisol levels to remain elevated even when they should be suppressed.

The distinction between acute stress, which is generally adaptive, and chronic stress, which is often harmful, is crucial. Acute stress responses evolved to deal with short-term physical challenges like escaping predators or fighting rivals. Once the challenge passes, stress systems are supposed to shut down, allowing recovery. Chronic stress, in contrast, involves repeated or sustained activation of stress systems without adequate recovery periods. The physiological changes that are adaptive in the short term become damaging when sustained.

How Chronic Stress Damages the Body

The health consequences of chronic stress are staggering in their scope and severity. Research over the past several decades has established that chronic psychological stress contributes to cardiovascular disease, immune dysfunction, metabolic disorders, gastrointestinal problems, chronic pain, accelerated aging, and premature death. Understanding the biological mechanisms linking stress to disease reveals why managing stress is not merely about feeling better but about protecting fundamental health.

The cardiovascular system bears much of the burden of chronic stress. Repeated sympathetic activation keeps blood pressure elevated, heart rate increased, and blood vessels constricted. Over time, these hemodynamic changes damage the endothelium, the delicate inner lining of blood vessels. Damaged endothelium becomes inflamed and loses its normal ability to regulate vascular tone and prevent blood clotting. This sets the stage for atherosclerosis, the formation of fatty plaques in artery walls that narrow vessels and restrict blood flow.

Cortisol contributes to cardiovascular damage through multiple mechanisms. It promotes accumulation of visceral fat, the deep abdominal fat that surrounds internal organs and is particularly harmful metabolically. Visceral fat secretes inflammatory molecules called cytokines that promote inflammation throughout the cardiovascular system. Cortisol also affects lipid metabolism, typically increasing levels of low-density lipoprotein cholesterol, the "bad" cholesterol that contributes to plaque formation, while decreasing high-density lip protein cholesterol, the "good" cholesterol that helps remove cholesterol from arteries.

Large-scale epidemiological studies have confirmed the cardiovascular consequences of chronic stress. People experiencing high work stress, particularly the combination of high job demands and low control called job strain, show increased risk of hypertension and coronary heart disease. Caregivers for family members with dementia or other serious illnesses, who experience chronic stress that can continue for years, show accelerated progression of atherosclerosis and increased cardiovascular mortality. People experiencing chronic financial strain, housing insecurity, or discrimination show elevated cardiovascular disease risk even after controlling for traditional risk factors like smoking, obesity, and family history.

The immune system, tasked with defending against infections and cancer, is profoundly affected by chronic stress. This might seem paradoxical given that cortisol is widely used medically as an anti-inflammatory and immunosuppressive drug. How can stress both suppress immune function and promote inflammation? The answer lies in the complexity of immune regulation and the different effects of acute versus chronic stress.

Acute stress actually enhances certain aspects of immune function, particularly innate immunity, the first line of defense against pathogens. The sympathetic nervous system and stress hormones mobilize immune cells, preparing the body for potential infections that might result from injuries sustained during the threatening situation. This acute immune enhancement makes evolutionary sense as preparation for dealing with wounds.

However, chronic stress impairs immune function through multiple mechanisms. Prolonged cortisol elevation suppresses production and function of lymphocytes, the white blood cells crucial for adaptive immunity that targets specific pathogens. It impairs antibody production, reducing responses to vaccines and increasing susceptibility to infections. Studies by Sheldon Cohen and colleagues have elegantly demonstrated that people experiencing chronic stress are more vulnerable to developing colds when exposed to cold viruses, with stress predicting both whether people become infected and the severity of symptoms.

Paradoxically, chronic stress also promotes inflammation despite cortisol's anti-inflammatory effects. This occurs partly through glucocorticoid resistance, where immune cells become less responsive to cortisol's anti-inflammatory signals. Cells that normally reduce inflammatory responses when cortisol binds to their receptors fail to do so when receptors are down regulated or desensitized. This allows inflammatory processes to proceed unchecked despite high cortisol levels. Additionally, stress activates inflammatory pathways through mechanisms independent of cortisol, including sympathetic nervous system effects on immune cells.

The resulting chronic low-grade inflammation contributes to numerous diseases beyond infections. Inflammation plays central roles in cardiovascular disease, promoting plaque formation and rupture. It contributes to insulin resistance and type 2 diabetes. It is implicated in neurodegenerative diseases including Alzheimer's disease. It even contributes to depression, with inflammatory cytokines affecting brain function and neurotransmitter systems in ways that promote depressive symptoms. The recognition that stress promotes inflammation has revolutionized understanding of how psychological experiences affect physical health.

Metabolic consequences of chronic stress extend beyond cardiovascular effects. Sustained cortisol elevation promotes insulin resistance, where cells become less responsive to insulin's signals to take up glucose from the bloodstream. This forces the pancreas to produce more insulin to achieve the same glucose-lowering effects. Over time, the pancreas may become unable to keep up with increased insulin demands, leading to elevated blood glucose and eventually type 2 diabetes. Chronic stress is now recognized as a significant risk factor for diabetes, with stressed individuals showing accelerated progression from normal glucose regulation to prediabetes to diabetes.

Stress also affects eating behavior and metabolism in ways that promote weight gain, particularly accumulation of harmful visceral fat. Some people eat less when stressed, but many eat more, particularly high-calorie comfort foods rich in sugar and fat. These eating changes appear to be mediated partly by stress effects on reward systems in the brain, with stressed individuals showing enhanced response to food rewards. Cortisol directly promotes visceral fat accumulation through effects on fat cell receptors that are particularly dense in the abdominal region. The resulting obesity further increases health risks through its own effects on cardiovascular, metabolic, and immune function.

Gastrointestinal problems are among the most commonly reported stress-related symptoms. The gut is often called the "second brain" because it contains extensive neural networks and produces many of the same neurotransmitters found in the brain. Stress affects gut function through multiple pathways including the sympathetic nervous system, which inhibits digestion during acute stress, and the HPA axis, which alters gut motility, secretion, and barrier function.

Chronic stress is associated with numerous gastrointestinal disorders including irritable bowel syndrome, inflammatory bowel disease, peptic ulcers, and functional dyspepsia. The mechanisms are complex and bidirectional. Stress affects the gut microbiome, the trillions of bacteria living in the intestines that influence digestion, immune function, and even brain chemistry. Disrupted microbiome composition, called dysbiosis, contributes to gastrointestinal symptoms and may feed back to affect stress responses through the gut-brain axis. Stress also increases intestinal permeability, allowing bacteria and their products to cross the intestinal barrier and trigger inflammatory responses.

Chronic pain conditions including headaches, back pain, fibromyalgia, and temporomandibular joint disorders are strongly associated with chronic stress.

Stress increases muscle tension throughout the body, particularly in the neck, shoulders, and jaw, contributing to tension headaches and muscle pain. Stress also affects pain perception through alterations in brain regions that process pain signals, typically increasing pain sensitivity. The relationship between stress and pain is bidirectional, with pain itself being a stressors that can perpetuate stress responses, creating vicious cycles where stress increases pain and pain increases stress.

Perhaps most concerning, chronic stress accelerates biological aging at the cellular level. Research by Elizabeth Blackburn, who won the Nobel Prize for discovering telomerase, has shown that chronic stress is associated with shortened telomeres, protective caps on the ends of chromosomes that normally shorten with each cell division. When telomeres become critically short, cells can no longer divide and may die or become senescent, contributing to aging and age-related diseases. Chronically stressed individuals, particularly those experiencing severe stressors like caring for chronically ill children, show accelerated telomere shortening equivalent to years of additional aging.

The cumulative burden of these stress-related physiological changes has been captured in the concept of allostatic load, proposed by neuroendocrinologist Bruce McEwen. Allostasis refers to achieving stability through change, the normal process of adapting to challenges by altering physiological set points. Allostatic load refers to the cumulative wear and tear on the body resulting from repeated cycles of allostasis, particularly when demands exceed adaptive capacity or when stress response systems fail to shut down after challenges pass. High allostatic load, measured through biomarkers including cortisol, blood pressure, glucose, lipids, and inflammatory markers, predicts increased risk of disease and mortality.

Stress and the Brain

Beyond its effects on the body, chronic stress profoundly affects brain structure and function in ways that impair cognition, alter emotional processing, and increase vulnerability to mental illness. The brain regions most affected by stress are ironically those most involved in regulating stress responses, creating potential for self-perpetuating cycles where stress-induced brain changes make individuals less able to cope with future stress.

The hippocampus, crucial for memory and for providing contextual information that helps regulate stress responses, is particularly vulnerable to chronic stress. Animal studies have consistently shown that chronic stress causes hippocampal neurons to atrophy, with dendrites retracting and losing synaptic connections. Chronic stress also suppresses neurogenesis, the birth of new neurons in the hippocampus that continues throughout life and contributes to learning and emotional regulation. In humans, brain imaging studies show reduced hippocampal volume in people experiencing chronic stress, including combat veterans with PTSD, people with chronic depression, and victims of childhood abuse.

These structural changes have functional consequences. Reduced hippocampal function impairs several forms of memory, particularly memory for spatial information and episodic memories for events in their temporal and spatial context. It may also impair the hippocampus's ability to regulate HPA axis responses by providing contextual information about whether situations are truly threatening. This creates potential for overgeneralization of stress responses, where individuals respond to safe situations as if they were dangerous because they cannot effectively distinguish contexts.

The prefrontal cortex, particularly the medial and orbitofrontal regions involved in emotion regulation, decision-making, and behavioral control, also shows structural and functional changes with chronic stress. Neurons in these regions show dendritic atrophy similar to that seen in the hippocampus.

Functional connectivity between prefrontal regions and the amygdala often decreases, potentially impairing top-down regulation of emotional responses. These prefrontal changes may contribute to the poor decision-making, impaired planning, and difficulty regulating emotions often seen in chronically stressed individuals.

In contrast to shrinkage in hippocampus and prefrontal cortex, the amygdala often shows increased volume and heightened reactivity with chronic stress. The amygdala's neurons show dendritic growth rather than atrophy, expanding their capacity to receive inputs and form connections. This structural expansion coincides with behavioral evidence of heightened anxiety and threat sensitivity. The combination of enhanced amygdala function with reduced hippocampal and prefrontal function creates a brain biased toward detecting threats, less able to regulate emotional responses, and less able to learn that previously dangerous situations have become safe.

At the neurochemical level, chronic stress affects multiple neurotransmitter systems. Serotonin function is often impaired, contributing to depressed mood, anxiety, and poor impulse control. Norepinephrine systems may become dysregulated, sometimes showing excessive reactivity to stressors. Dopamine function in reward circuits may be blunted, reducing the ability to experience pleasure and motivation, a state called anhedonia that is central to depression. The combination of altered neurotransmitter function with structural brain changes helps explain why chronic stress is a major risk factor for depression, anxiety disorders, and substance abuse.

Importantly, many stress-induced brain changes are reversible if stress is reduced or coping improves. Animal studies show that removing stressors allows hippocampal neurons to regrow dendrites and resume normal neurogenesis. In humans, interventions including stress management therapy, meditation, and exercise can increase hippocampal volume and improve prefrontal function. This reversibility provides hope that stress damage need not be permanent and that intervention can promote neural recovery.

Individual Differences in Stress Vulnerability

A striking feature of stress responses is that people vary enormously in how they respond to similar stressors. Some individuals develop stress-related illness after relatively minor challenges, while others remain healthy despite enduring extraordinary adversity. Understanding sources of this variation has important implications for identifying who needs intervention and what interventions might be most helpful.

Genetic factors contribute significantly to variation in stress vulnerability. Twin studies suggest that roughly thirty to fifty percent of variation in stress responses and stress-related disorders like depression and anxiety has genetic components. However, stress vulnerability is not determined by single genes but rather reflects combined influences of many genes, each with small individual effects, interacting with environmental factors.

Some of the genes involved in stress vulnerability affect the HPA axis and glucocorticoid receptors. Variations in genes encoding glucocorticoid receptors influence how effectively cortisol provides negative feedback to shut down stress responses. People with less effective feedback may show prolonged cortisol elevation after stress. Other genes affect enzymes that synthesize or break down cortisol, influencing how much cortisol exposure tissues experience. Still others affect proteins that transport cortisol or regulate glucocorticoid receptor function.

Genes affecting neurotransmitter systems also influence stress vulnerability. The serotonin transporter gene, which has a common variant affecting how efficiently serotonin is removed from synapses, has been extensively studied. Some research suggests that people carrying the less efficient variant show greater stress reactivity and higher rates of depression following stressful life events, though these findings remain controversial and not all studies replicate them. Similar gene-by-environment interactions have been reported for genes affecting dopamine, norepinephrine, and other neurotransmitter systems.

An important insight from genetic research is that genes affecting stress vulnerability are not simply "bad" genes that create problems. Many appear to create differential susceptibility to environmental influences, making carriers more vulnerable to adverse environments but also more responsive to positive environments. This has been called the orchid-dandelion hypothesis, suggesting that some people are like hardy dandelions who do reasonably well in any environment, while others are like delicate orchids who wither in adverse environments but flourish more than dandelions when conditions are favorable. This re frames genetic "vulnerability" as genetic sensitivity to experience.

Early life experiences profoundly shape stress response systems in ways that can last a lifetime. Adverse childhood experiences, including abuse, neglect, parental loss, witnessing violence, and household dysfunction, are among the strongest predictors of stress-related health problems in adulthood. The landmark Adverse Childhood Experiences (ACE) study, conducted by the Centers for Disease Control and Kaiser Permanente, found that the number of adverse experiences in childhood predicts adult health in a dose-response manner, with more adversities predicting higher rates of depression, anxiety, substance abuse, cardiovascular disease, diabetes, and premature death.

The biological mechanisms linking early adversity to later health involve programming of stress response systems during sensitive periods of development. Animal research demonstrates that the quality of maternal care during early life permanently alters offspring's stress responses through epigenetic mechanisms. Rat pups receiving high levels of maternal care through licking and grooming develop HPA axes that respond more dynamically to stress, with sharp increases followed by efficient returns to baseline. Pups receiving low maternal care develop blunted or dysregulated stress responses that persist into adulthood.

Epigenetic modifications, changes in gene expression without changes in DNA sequence, provide mechanisms for lasting effects of early experience. Early-life stress can alter methylation of genes encoding glucocorticoid receptors in the hippocampus, reducing their expression and impairing negative feedback regulation of cortisol. These epigenetic changes can persist throughout life, creating lasting alterations in stress biology. Human studies have found similar epigenetic modifications associated with childhood maltreatment, suggesting that early adversity may "get under the skin" through these molecular mechanisms.

However, early adversity does not rigidly determine outcomes. Many people prove remarkably resilient despite significant childhood stress, and interventions can mitigate effects of early adversity. Factors promoting resilience include supportive relationships with other adults who provide care and stability, development of competencies and coping skills, and engagement in meaningful activities that build self-efficacy. Programs addressing childhood adversity increasingly focus on building these protective factors rather than simply trying to eliminate risk factors.

Personality traits and cognitive styles also influence stress vulnerability. Neuroticism, the tendency toward negative emotionality, worry, and distress, is one of the strongest personality predictors of stress-related health problems. People high in neuroticism perceive more situations as stressful, react more intensely to stressors, and recover more slowly. They also show heightened physiological stress reactivity including exaggerated cortisol responses. Conversely, personality traits including optimism, conscientiousness, and extroversion are associated with better stress outcomes.

Coping styles, the characteristic ways people respond to stress, powerfully influence outcomes. Problem-focused coping, which involves active efforts to change or manage stressful situations, generally produces better outcomes than emotion-focused coping centered on managing emotional distress without addressing underlying problems. However, coping effectiveness depends on situational controllability. Problem-focused coping works well for controllable stressors but can be frustrating and ineffective for uncontrollable stressors, where emotion-focused coping and acceptance may be more adaptive.

Perceived control, the belief that one has influence over outcomes, is among the most powerful psychological buffers against stress effects. Animal research demonstrates dramatically different physiological consequences of controllable versus uncontrollable stress. Rats exposed to electric shocks they can terminate by performing a behavior show minimal stress-related pathology compared to rats receiving identical shocks that they cannot control. The physical stressors is identical; the difference in controllability creates vastly different biological and behavioral consequences.

In humans, perceived control predicts stress outcomes across diverse domains. Workers experiencing high demands paired with low control show particularly elevated disease risk in the job strain model mentioned earlier. Nursing home residents given more control over daily decisions show better health and longer survival than those with equivalent care but less control. People who believe they have control over stressful situations show smaller cortisol responses and less stress-related illness even when objective controllability is similar.

Social support represents another powerful buffer against stress. People with strong social connections show reduced physiological stress responses, better immune function, lower rates of mental and physical illness, and lower mortality rates compared to socially isolated individuals. Social support appears to buffer stress through multiple mechanisms including provision of practical assistance with problems, emotional comfort during distress, and cognitive re framing that helps people view situations as more manageable.

The biological mechanisms underlying social support effects include the neuropeptide oxytocin, which is released during positive social interactions and reduces HPA axis activation, decreases amygdala reactivity, and promotes parasympathetic calming. Social support may also work through reducing inflammatory responses to stress, with supported individuals showing smaller increases in inflammatory cytokines following stressors. Some research suggests that thinking about supportive relationships or viewing photographs of loved ones can reduce stress responses, suggesting that mental representations of support confer benefits even without active social interaction.

However, not all social relationships reduce stress. Conflictual relationships, unsupportive responses to stress disclosure, and social isolation all increase stress and health risks. The quality rather than mere quantity of relationships appears crucial, with one close, supportive relationship providing more benefit than numerous superficial connections. Understanding the protective mechanisms of social support has led to interventions aimed at strengthening social ties as part of stress management programs.

Stress Management: Taking Control of Your Biology

Understanding the biology of stress empowers people to take active steps to manage stress responses and protect health. While some stressors cannot be eliminated, how the body and brain respond to stressors can be modified through interventions targeting multiple levels from behavior to neurobiology. The most effective stress management approaches combine multiple strategies tailored to individual needs and circumstances.

Physical exercise is one of the most powerful stress management tools, with effects on both acute stress responses and long-term stress resilience. Acute exercise temporarily increases cortisol and sympathetic activity, similar to stress. However, regular exercise training produces adaptations that reduce stress responses. Physically fit individuals show smaller cortisol and cardiovascular responses to psychological stressors compared to unfit individuals. Exercise also promotes neurogenesis in the hippocampus, increases brain-derived neurotrophic factor that supports neural health, and reduces inflammation.

The stress-buffering effects of exercise appear strongest for moderate-intensity activity performed regularly rather than infrequent intense exercise. Activities combining physical exertion with enjoyment and social interaction may provide additional benefits through multiple mechanisms. The type of exercise matters less than consistency and sustainability, suggesting that the best exercise for stress management is whatever activity a person will actually maintain long-term.

Relaxation techniques including progressive muscle relaxation, deep breathing, and autogenic training produce physiological states opposite to stress activation. These practices engage the parasympathetic nervous system, slowing heart rate, lowering blood pressure, reducing muscle tension, and promoting calm. Regular practice appears to reduce baseline physiological arousal and improve ability to activate relaxation responses when needed. Some research suggests that relaxation training can reduce cortisol levels, improve immune function, and decrease inflammatory markers.

Mindfulness meditation, which involves non-judgmental attention to present-moment experience, has attracted enormous research interest as a stress management intervention. Studies suggest that mindfulness training reduces perceived stress, anxiety, and depression while improving emotion regulation and wellbeing. Brain imaging studies show that mindfulness practice increases activity and gray matter volume in prefrontal regions involved in attention and emotion regulation while reducing amygdala reactivity. Long-term meditation practitioners show reduced cortisol responses to stress and healthier patterns of cortisol secretion throughout the day.

The mechanisms through which mindfulness reduces stress likely involve both cognitive and physiological pathways. Cognitively, mindfulness may reduce stress by changing relationships to thoughts and feelings, fostering acceptance rather than struggle with difficult experiences. Physiologically, meditation appears to enhance prefrontal regulation of limbic regions, reducing automatic stress reactivity. However, mindfulness is not a panacea, and effects vary considerably across individuals. Some people benefit greatly while others find meditation difficult or unhelpful, suggesting the need for personalized approaches.

Cognitive-behavioral stress management targets appraisal and coping processes that mediate relationships between stressors and stress responses. These interventions teach skills for identifying and challenging stress-amplifying thoughts, developing more adaptive appraisals, improving problem-solving abilities, and expanding coping repertoires. By changing how situations are interpreted and what coping resources are perceived as available, cognitive approaches can reduce stress responses even when objective stressors remain unchanged.

Research shows that cognitive-behavioral interventions can reduce physiological stress markers including cortisol and inflammatory cytokines, improve immune function, and reduce stress-related illness. The effects appear mediated partly through reduced amygdala reactivity and enhanced prefrontal regulation observed in brain imaging studies. Cognitive approaches may be particularly effective for chronic stressors that cannot be eliminated but can be re framed or approached differently.

Sleep is both affected by stress and a crucial buffer against stress effects. Chronic stress disrupts sleep through sustained physiological arousal that interferes with sleep initiation and maintenance, and through worry that prevents mental quieting needed for sleep. Insufficient or poor-quality sleep then impairs stress coping and exacerbates stress responses, creating vicious cycles. Interventions that improve sleep quality appear to reduce stress vulnerability, with some research suggesting that resolving sleep problems can reduce depression and anxiety even without directly targeting those conditions.

Nutrition interacts with stress in complex ways. Stress affects eating behavior, often increasing consumption of high-calorie comfort foods that provide temporary mood benefits through effects on reward circuits and possibly through stress-dampening effects of certain nutrients. However, stress eating typically involves foods high in sugar, fat, and calories that contribute to obesity and metabolic problems. Conversely, some evidence suggests that consuming a healthy diet rich in fruits, vegetables, whole grains, and omega-3 fatty acids may buffer stress effects, though mechanisms remain unclear and research findings are mixed.

Time in nature appears to reduce stress through multiple pathways. Natural environments promote parasympathetic activation and reduce cortisol more effectively than urban environments. Even brief exposures to nature, including viewing nature through windows or looking at nature photographs, can reduce stress responses. The mechanisms may involve both psychological effects of natural beauty and reduced sensory demands compared to complex urban environments. Some research suggests that forest environments may release airborne chemicals that affect immune function, though this remains speculative.

Social connection deserves special emphasis as perhaps the most powerful stress buffer. Maintaining supportive relationships, seeking social support during stressful periods, and contributing to others' well being all appear to reduce stress effects on health. However, modern life often isolates people, with many reporting insufficient social connection despite being constantly connected through digital media. The quality of face-to-face interaction appears particularly important, with digital connection providing weaker benefits than in-person social contact.

For severe chronic stress or stress-related disorders, professional help through therapy or medication may be necessary. Cognitive-behavioral therapy, mindfulness-based interventions, and other evidence-based psychotherapies can provide structured approaches to stress management tailored to individual needs. Medications including antidepressants may be helpful when stress has contributed to clinical depression or anxiety disorders, though medication alone without behavioral change typically provides incomplete benefit.

Stress as Growth: The Potential for Stress-Related Growth

While most discussion of stress focuses on its harmful effects, emerging research explores positive changes that can follow from successfully navigating stressful experiences. The concept of post-traumatic growth describes positive psychological changes including increased personal strength, closer relationships, greater appreciation for life, new possibilities or life paths, and spiritual development that some people report following traumatic events.

This does not mean that trauma or extreme stress are good or desirable. Most people experiencing severe stress would trade their growth for not having faced the stressor. However, recognizing that positive change can emerge from adversity may help people find meaning in suffering and identify opportunities for growth. Research suggests that moderate levels of lifetime stress exposure, compared to either very low or very high exposure, may promote resilience by providing opportunities to develop coping skills without overwhelming adaptive capacity.

The biology of stress-related growth likely involves neuroplasticity and learning processes. Successfully coping with challenges may strengthen neural circuits involved in emotion regulation and problem-solving, creating lasting enhancements in coping capacity. Overcoming fears through exposure may strengthen extinction circuits, reducing anxiety in future threatening situations. Social support mobilized during crises may strengthen relationships in lasting ways. These biological and psychological changes can leave people better equipped to handle future stress.

Understanding stress as a potentially growth-promoting challenge rather than simply a threat may itself reduce stress effects. Research on stress mindsets suggests that viewing stress as enhancing rather than debilitating can reduce negative health effects and improve performance. This does not mean denying that stress is difficult or pretending that severe stress is beneficial. Rather, it involves recognizing that the body's stress response evolved to help meet challenges and that moderate stress can enhance learning, performance, and growth when coped with successfully.

Conclusion: Living with Stress in the Modern World

Stress is an unavoidable aspect of human life, rooted in biological systems that evolved to help our ancestors survive in dangerous environments. These same systems continue to serve important protective functions today, preparing us to meet challenges and recover from adversity. However, the chronic psychological stressors of modern life often activate these systems in ways they were not designed to handle, creating enormous burdens of stress-related illness.

Understanding the biology of stress reveals both the power of these ancient systems and the possibility of managing them more effectively. Stress is not simply an inevitable consequence of circumstances but rather a biological response shaped by how situations are appraised, what coping resources are available, and how stress response systems have been programmed by genes and experience. This understanding empowers people to take active steps to manage stress through behavioral, cognitive, and social interventions that can alter biology as surely as medications do.

The goal is not to eliminate all stress, which would be impossible and perhaps undesirable. Moderate, controllable stress promotes learning, growth, and resilience. The goal instead is to reduce chronic, uncontrollable stress while building capacity to cope effectively with unavoidable challenges. This requires both individual action and social change, as many of the most harmful stressors including poverty, discrimination, and social isolation require collective solutions.

8

LOVE, BONDING, AND SOCIAL CONNECTION

Humans are profoundly social creatures. From the moment of birth, when a newborn instinctively seeks the warmth and nourishment of a caregiver, to the final moments of life, when people overwhelmingly wish to be surrounded by loved ones, our need for connection with others shapes every aspect of existence. We form attachments to parents and children, romantic bonds with partners, friendships with peers, and connections to communities and cultures. These social bonds are not mere luxuries or pleasant additions to life but fundamental biological needs as essential as food, water, and shelter.

The depth of this social need becomes starkly apparent when it is absent. Infants deprived of physical contact and emotional connection fail to thrive even when their physical needs for nutrition and hygiene are met, sometimes developing severe developmental delays or even dying from the lack of human touch and interaction. Adults subjected to prolonged social isolation experience profound psychological distress, cognitive decline, and increased mortality risk. Loneliness has been called an epidemic in modern societies, with health consequences comparable to smoking or obesity. These observations reveal that social connection is not simply psychologically comforting but biologically necessary.

The scientific study of love and attachment was revolutionized in the mid-twentieth century by psychologist Harry Harlow's landmark experiments with infant rhesus monkeys. Harlow created two surrogate mothers for orphaned infant monkeys: one made of bare wire that provided milk through an attached bottle, and another covered in soft terry cloth that provided no food. According to prevailing theories of the time, which emphasized feeding as the basis of attachment, infants should have preferred the wire mother who provided nutrition. Instead, infant monkeys spent most of their time clinging to the cloth mother, visiting the wire mother only briefly when hungry before returning to the comfort of soft contact.

These experiments demonstrated that attachment is not simply about obtaining food or other material resources but involves a fundamental need for comfort, security, and physical contact. Harlow's work was controversial and would not meet modern ethical standards for animal research, but it fundamentally changed understanding of what drives social bonding and laid groundwork for studying the neurobiology of attachment and love.

This chapter explores the biological foundations of human social connection, examining the brain systems that motivate us to seek relationships, the hormones that create feelings of bonding and trust, how early attachment experiences shape development, the neuroscience of romantic love, why social isolation damages health, and how understanding the biology of connection can help us build healthier relationships and communities.

The Neurobiology of Attachment: Bonds That Begin in Infancy

The formation of attachment bonds between infants and caregivers represents perhaps the most crucial social relationship in human life, laying foundations for emotional development, stress regulation, and future relationships. Attachment theory, developed by psychiatrist John Bowl by and psychologist

Mary Ainsworth, proposes that infants are biologically predisposed to form strong emotional bonds with caregivers who provide safety and security. These bonds serve obvious survival functions, keeping vulnerable infants close to adults who can protect and nurture them.

Bowl by recognized that attachment behaviors including crying, clinging, following, and smiling are not learned through rewards and punishments but rather are innate responses shaped by evolution. Infants who formed strong attachments to caregivers were more likely to receive protection and care, increasing their chances of survival and reproduction. Natural selection therefore favored genetic variations that promoted strong attachment bonds. This evolutionary perspective explains why attachment behaviors appear universally across cultures and why disruptions to attachment are so distressing.

Ainsworth's research identified different patterns of infant attachment using an experimental procedure called the Strange Situation. In this procedure, mothers and infants are observed during a series of brief separations and reunions in an unfamiliar room, sometimes with a stranger present. Infants' responses to these mildly stressful separations reveal their attachment security.

Securely attached infants, comprising about sixty to sixty-five percent of middle-class samples in Western cultures, use their caregivers as a secure base for exploration when together, show distress when separated, and seek comfort when reunited. They are easily soothed by the caregiver's return and then resume exploration. These patterns reflect trust that the caregiver will be available and responsive when needed.

Insecurely attached infants show different patterns. Avoidant infants, about twenty percent of samples, show little distress during separation and avoid or ignore the caregiver upon reunion, appearing emotionally distant. Resistant or ambivalent infants, about ten to fifteen percent, show intense distress during separation but are difficult to comfort upon reunion, sometimes showing anger or continued distress even while seeking proximity. Disorganized infants, about five to ten percent but higher in high-risk populations, show confused or contradictory behaviors suggesting they lack coherent strategies for dealing with stress.

These attachment patterns reflect not infant temperament but rather the quality of care giving received. Sensitive, responsive care giving that reliably meets infant needs promotes secure attachment. Inconsistent, insensitive, or rejecting care giving promotes insecure patterns. Frightening or traumatizing care giving promotes disorganized attachment. Importantly, attachment security can change if care giving quality changes, and secure attachments can be promoted through interventions that improve caregiver sensitivity and responsiveness.

Neuroscience research has begun to identify the brain systems underlying attachment bonds. The neuropeptide oxytocin plays particularly crucial roles in both maternal caregiving and infant attachment. Oxytocin is released in mothers during childbirth and breastfeeding, promoting maternal behavior and bonding with infants. Mothers with higher oxytocin levels show more affectionate behavior toward infants, more synchrony in interactions, and stronger activation of reward circuits when viewing their own infants.

Oxytocin also affects infant attachment, though research in humans is more limited for ethical reasons. Animal studies show that blocking oxytocin receptors disrupts attachment behaviors in infant animals, while enhancing oxytocin signaling strengthens bonds. In humans, genetic variations affecting oxytocin receptors are associated with differences in attachment security, suggesting oxytocin contributes to individual differences in bonding.

The brain's reward circuits, centered on dopamine systems originating in the ventral tegmental area and projecting to the nucleus accumbens, appear crucial for attachment motivation. Brain imaging studies show that mothers viewing pictures of their own infants show strong activation in these reward regions, similar to activation seen when people view romantic partners or receive other rewarding stimuli. This suggests that infants are literally rewarding to parents, creating positive feelings that motivate continued caregiving and bonding.

The attachment system also involves brain regions regulating stress and emotion, particularly the amygdala and hypothalamus. Securely attached infants show more moderate stress responses, measured by cortisol levels, during separations compared to insecurely attached infants. The presence of attachment figures appears to buffer stress responses, with infants showing smaller stress reactions when caregivers are present than when alone. This stress-buffering function of attachment relationships continues throughout life, with supportive relationships reducing stress responses in children, adolescents, and adults.

Early attachment experiences appear to program stress response systems in lasting ways. Animal research demonstrates that maternal care quality during early life alters offspring stress biology through epigenetic mechanisms, as discussed in the previous chapter. Rat pups receiving high maternal care develop HPA axes that respond dynamically to stress and then efficiently return to baseline. Pups receiving low maternal care show dysregulated stress responses that persist into adulthood and are transmitted to the next generation through altered maternal behavior.

In humans, secure attachment is associated with healthier stress regulation, better emotion regulation abilities, and lower rates of anxiety and depression throughout life. Insecure and particularly disorganized attachment predict increased risk for mental health problems, relationship difficulties, and stress-related physical illness. These associations suggest that early attachment experiences create biological vulnerabilities or resilience that influence health across the lifespan. However, attachment patterns are not destiny, and later positive relationships can partly compensate for insecure early attachments.

Understanding attachment as a biological system with roots in evolution and neurobiology helps explain why disruptions to early caregiving have such profound consequences. Institutionalized children who experience severe deprivation of consistent caregiving show delays in physical growth, cognitive development, and emotional regulation. Many show persistent difficulties with relationships and emotion regulation even after adoption into nurturing families, though earlier adoption generally produces better outcomes than later adoption.

These findings have transformed childcare practices and policies. Recognition that infants need consistent, responsive caregiving has led to reforms in orphanages and foster care systems emphasizing stable placements with trained caregivers. Parenting interventions increasingly focus on promoting sensitive, responsive caregiving as the foundation for healthy development. The biological understanding of attachment provides compelling rationale for investing in early childhood relationships and supporting parents in providing the nurturing care that programs healthy development.

Oxytocin: The Chemistry of Bonding

Among the many neurochemicals involved in social bonding, oxytocin has received particular attention as a hormone that promotes trust, empathy, and social connection. Oxytocin is a peptide hormone and neuromodulator synthesized in the hypothalamus and released into the bloodstream by the posterior pituitary gland. It is also released within the brain where it acts as a neurotransmitter, binding to oxytocin receptors on neurons throughout the nervous system.

Oxytocin was first recognized for its roles in reproduction, particularly in stimulating uterine contractions during childbirth and milk ejection during breastfeeding. The name oxytocin derives from Greek words meaning "quick birth," reflecting its role in labor. Synthetic oxytocin is routinely used medically to induce or augment labor. However, research over the past few decades has revealed that oxytocin's functions extend far beyond reproduction to include profound effects on social behavior, emotion, and stress regulation.

The discovery that oxytocin promotes social bonding came partly from studies of prairie voles, small rodents that form monogamous pair bonds between mating partners, unusual among mammals. When prairie voles mate, oxytocin and vasopressin, a closely related neuropeptide, are released in the brain. This hormonal release appears to create the partner preference that characterizes vole pair bonds. Blocking oxytocin receptors prevents pair bond formation even after mating, while artificially activating oxytocin receptors can create partner preferences even without mating. These findings suggested that oxytocin might be a critical neurochemical basis of social bonding.

In humans, oxytocin levels rise during positive social interactions including conversation, cooperation, and particularly physical touch. Hugging, massage, and sexual activity all trigger oxytocin release. The release of oxytocin during these interactions appears to create positive feelings associated with social connection and motivates seeking further contact. This creates positive feedback loops where social interaction releases oxytocin, which makes interaction feel rewarding, which motivates more interaction, which releases more oxytocin.

Experimental studies have examined oxytocin effects by administering synthetic oxytocin as a nasal spray that crosses from the nasal cavity into the brain. These studies suggest that oxytocin increases trust, generosity, and cooperation in economic games, enhances ability to read emotions in faces, increases eye contact and social approach, and reduces anxiety in social situations. However, effects appear context-dependent and complex rather than simply making people more social in all circumstances.

Some research suggests oxytocin promotes trust and cooperation primarily toward in-group members while potentially increasing defensive responses toward out-group members, suggesting it may enhance tribal psychology rather than universal social harmony. Oxytocin may also increase sensitivity to social information generally rather than specifically promoting positive sociality, making people more responsive to both positive and negative social cues. These complexities have tempered initial enthusiasm that oxytocin might be a simple "love hormone" that could be used to enhance social functioning.

Nevertheless, the evidence that oxytocin plays important roles in human bonding remains compelling. Brain imaging studies show that oxytocin modulates activity in regions involved in social cognition and emotion, particularly the amygdala. Oxytocin typically reduces amygdala responses to threatening social stimuli while enhancing responses to positive social signals. Oxytocin also appears to enhance connectivity between the amygdala and prefrontal regions involved in emotion regulation, potentially improving ability to regulate social anxiety.

Genetic variations affecting oxytocin receptors are associated with differences in social behavior and relationship quality. People carrying certain variants show differences in attachment security, empathy, and stress reactivity. However, these genetic effects are modest and interact with environmental factors, consistent with gene-environment interactions discussed in earlier chapters.

Clinical research has explored oxytocin as a potential treatment for conditions involving social difficulties, particularly autism spectrum disorders. Some studies suggest that oxytocin administration may temporarily improve social communication and reduce anxiety in individuals with autism, though effects are variable and more research is needed to determine whether oxytocin has lasting therapeutic benefits. The complexity of oxytocin's effects and questions about optimal dosing and administration make clinical application challenging.

Beyond its effects on bonding, oxytocin powerfully influences stress responses. As mentioned in the previous chapter, oxytocin reduces HPA axis activation and promotes parasympathetic nervous system activity, creating physiological calm. This stress-buffering function may be one mechanism through which social support protects health. Positive social interactions release oxytocin, which reduces stress responses both in the moment and potentially with cumulative effects over time in people with strong social connections.

Oxytocin also interacts with the endogenous opioid system, the brain's natural pain-relieving and pleasure-producing neurochemical system. Social touch releases both oxytocin and endogenous opioids, creating feelings of comfort and pleasure that make social contact intrinsically rewarding. This may explain why physical touch from trusted others is so soothing during stress and why lack of touch is so distressing. The COVID-19 pandemic, which required prolonged social distancing and reduced physical contact, provided natural experiments revealing how much people suffer when deprived of touch and close social interaction.

Romantic Love: The Neuroscience of Passion and Commitment

Romantic love, one of the most intense and consequential human experiences, has inspired countless poems, songs, and works of art throughout history. The experience of falling in love involves obsessive thoughts about the beloved, intense longing for their presence, euphoria when together, and devastation during separation or rejection. These powerful emotions motivate extraordinary behavior including self-sacrifice, risk-taking, and radical life changes. Understanding the neurobiology of romantic love reveals both universal biological foundations and cultural variations in how love is experienced and expressed.

Psychologists distinguish between passionate love, the intense longing and physiological arousal characteristic of early romantic relationships, and companionate love, the deep affection and commitment that characterize long-term partnerships. These different forms of love appear to involve partially distinct neural systems, with passionate love showing more overlap with reward and motivation circuits while companionate love shows more overlap with attachment systems.

Brain imaging studies of people in the early stages of romantic love show strong activation of reward circuits when viewing photographs of their romantic partners. The ventral tegmental area, which contains dopamine neurons that drive reward and motivation, shows particularly robust activation. This region is the same one activated by addictive drugs, leading researchers to propose that romantic love functions like a natural addiction, creating intense craving for the beloved and withdrawal symptoms during separation.

The intensity of early romantic love and its similarity to addiction helps explain the dramatic behaviors often seen in new lovers. The dopamine-driven motivation to be near the beloved can override other goals and concerns. The agony of rejection or breakup resembles withdrawal from addictive drugs, with similar despair, craving, and sometimes desperate attempts to regain the relationship. The obsessive thinking about the beloved reflects the all-consuming nature of dopamine-driven motivation when focused on a particular goal.

However, characterizing romantic love as merely addiction oversimplifies. Love also involves activation of attachment systems, particularly regions including the anterior cingulate cortex and insula that process emotional states and generate feelings of connection. As relationships progress from passionate to companionate love, brain activation patterns shift. The intense dopamine-driven reward responses of early love typically diminish, while activation of attachment-related regions involved in calm, secure bonding increases.

Long-term romantic partners show activation patterns when viewing each other that combine elements of both reward and attachment systems, along with reduced activation in regions involved in fear and negative emotion. This suggests that mature love involves sustained positive feelings without the intense highs and lows of early passion. However, some couples maintain passionate love even in long-term relationships, showing that passionate and companion ate love need not be mutually exclusive.

The neurochemistry of romantic love involves multiple systems beyond dopamine. Oxytocin and vasopressin, the neuropeptides involved in attachment, appear crucial for the formation and maintenance of romantic bonds. Oxytocin levels rise during positive romantic interactions including conversation, affection, and sexual activity. Genetic variations affecting oxytocin and vasopressin receptors are associated with differences in relationship quality and stability, though effects are modest.

Serotonin appears to be reduced in early romantic love, a pattern similar to that seen in obsessive-compulsive disorder. This may contribute to the obsessive preoccupation with the beloved characteristic of early love. As relationships mature and serotonin levels normalize, obsessive thinking typically diminishes. The temporary reduction in serotonin during early romance might serve adaptive functions by focusing attention intensely on establishing bonds during the critical early period.

Testosterone appears to decrease in men during early romantic relationships while increasing in women, potentially reducing gender differences in relationship behavior. These hormonal shifts might promote behaviors conducive to bonding rather than competition or sexual conquest. However, these effects are subtle and variable, and the relationships between hormones and romantic behavior remain incompletely understood.

Sexual attraction and romantic love involve overlapping but distinguishable neural systems. Sexual desire activates hypothalamic regions involved in sexual motivation and arousal, along with reward circuits. Romantic love activates some of the same reward regions but also involves regions associated with higher-level emotional processing and goal-directed behavior. This distinction helps explain how sexual desire and romantic love can occur independently, with people sometimes experiencing intense sexual attraction without romantic feelings, or romantic feelings without strong sexual desire.

Heartbreak, the emotional pain following romantic rejection or separation, activates brain regions involved in physical pain including the anterior cingulate cortex and insula. This neural overlap between social and physical pain may explain why heartbreak is described using physical pain metaphors like "broken heart" and why rejection literally hurts. Understanding that social pain activates similar neural circuits as physical pain validates the intense suffering caused by romantic loss and helps explain why people sometimes turn to pain medications to cope with heartbreak.

The cross-cultural study of romantic love reveals both universals and variations. The capacity for romantic love appears universal, with anthropological studies finding evidence of passionate love in the vast majority of cultures studied. However, cultures vary enormously in attitudes toward romantic love, with some viewing it as the ideal basis for marriage while others consider it irrelevant or even harmful to mate selection. These cultural variations in attitudes shape how love is experienced and expressed without eliminating the underlying biological capacity.

Arranged marriages, common in many cultures, demonstrate that romantic love is not necessary for successful long-term partnerships. Many arranged marriages develop deep companion ate love over time, challenging Western assumptions that romantic attraction must precede commitment. This suggests that while romantic passion may help initiate relationships, companion ate love and commitment can develop through alternative pathways including shared experiences, interdependence, and deliberate cultivation of affection.

Social Connection and Physical Health

The recognition that social relationships profoundly influence physical health represents one of the most important discoveries in health psychology. Decades of research have established that social isolation and loneliness increase mortality risk comparable to well-established risk factors including smoking, obesity, and physical inactivity. Conversely, strong social connections protect health and extend lifespan through multiple biological pathways.

The landmark Alameda County Study, which followed thousands of California residents for decades beginning in the 1960s, provided early evidence of social relationships' health effects. People with fewer social connections showed higher mortality rates over nine years of follow-up even after controlling for health behaviors, socioeconomic status, and initial health status. Married people, those with close friends and relatives, and members of religious or other social groups all showed survival advantages compared to socially isolated individuals.

Subsequent research has consistently confirmed and extended these findings. A meta-analysis combining data from 148 studies including more than 300,000 participants found that people with stronger social relationships had fifty percent increased odds of survival over study follow-up periods compared to those with weaker relationships. The magnitude of this effect rivals or exceeds that of many medical interventions aimed at extending life.

The biological mechanisms linking social connection to health involve multiple pathways. Social support reduces stress responses, with supported individuals showing smaller cortisol increases, lower blood pressure responses, and reduced sympathetic activation during stressful situations compared to unsupported individuals. These acute stress-buffering effects accumulate over time, reducing wear and tear on physiological systems.

Social relationships also influence health behaviors that affect disease risk. Socially connected people are more likely to engage in healthy behaviors including exercise, healthy eating, adequate sleep, and medical care, while being less likely to smoke or abuse substances. These behavioral pathways account for some but not all of the health benefits of social connection, with biological mechanisms contributing independently.

The immune system is particularly responsive to social connection. Socially isolated individuals show altered immune function including reduced immune responses to vaccines, slower wound healing, increased inflammation, and possibly increased vulnerability to infections and cancer. Animal research provides clear causal evidence that social isolation impairs immunity. Rodents housed in social isolation show weakened immune responses compared to group-housed animals, with isolation effects reversible by returning animals to social housing.

The mechanisms linking social isolation to immune dysfunction involve stress pathways discussed in the previous chapter. Social isolation is stressful, activating the HPA axis and sympathetic nervous system in ways that alter immune function. Additionally, loneliness appears to trigger inflammatory gene expression programs that evolved to protect against bacterial infections likely to result from social confrontation and wounding. In modern contexts where social isolation rarely involves physical threats, this inflammatory response becomes maladaptive, contributing to chronic inflammation that promotes numerous diseases.

Cardiovascular health is also strongly influenced by social relationships. Socially isolated individuals show increased rates of hypertension, coronary artery disease, heart attack, and stroke. The mechanisms involve both stress pathways, with isolation increasing blood pressure and vascular damage, and behavioral pathways, with isolated individuals less likely to adhere to medical regimens. Some research suggests that loneliness predicts cardiovascular disease independent of other risk factors, highlighting the direct biological effects of social isolation.

Mental health is profoundly affected by social connection, with loneliness being one of the strongest risk factors for depression and anxiety. The relationship is bidirectional, with loneliness promoting mental illness and mental illness promoting social withdrawal, creating potential for vicious cycles. Social isolation also increases suicide risk, with socially connected individuals showing substantially lower rates across cultures and throughout history.

Cognitive function and dementia risk are influenced by social engagement. People with rich social lives show slower cognitive decline in aging and reduced dementia risk compared to socially isolated elders. The mechanisms may involve cognitive stimulation provided by social interaction, stress-buffering effects protecting the brain, or direct effects of social engagement on brain structure and function. Interventions promoting social engagement in older adults show promising effects on maintaining cognitive function, though more research is needed to determine optimal approaches.

Understanding the health importance of social connection has led to growing recognition of loneliness as a public health problem requiring intervention. Modern societies, particularly wealthy industrialized nations, show increasing rates of loneliness despite or perhaps because of technological connectivity. Social media and digital communication provide superficial contact but may substitute for the deep, meaningful relationships that provide health benefits. The COVID-19 pandemic dramatically highlighted the health consequences of isolation, with prolonged social distancing contributing to increases in depression, anxiety, and substance abuse.

Interventions to reduce loneliness and promote social connection include programs bringing together isolated individuals, training in social skills for those who struggle with relationships, addressing barriers to social participation including mobility limitations, and community development initiatives creating opportunities for meaningful interaction. Technology-based interventions show mixed results, with some programs successfully connecting isolated individuals while others may reinforce isolation by replacing in-person contact with digital substitutes.

The Dark Side of Relationships: When Connection Turns Toxic

While positive relationships promote health, negative relationships can be profoundly harmful. Relationship conflict, abuse, and toxic dynamics activate stress systems chronically and intensely, creating health risks that can exceed those of social isolation. Understanding both the benefits of positive relationships and the harms of negative relationships is essential for promoting health through social connection.

Marital conflict has been extensively studied as a stressor with measurable physiological consequences. Couples engaged in hostile interactions show elevated cortisol, blood pressure, and inflammatory markers, with effects persisting for hours or days after conflicts end. Chronic marital distress predicts increased risk for depression, cardiovascular disease, and slower wound healing. In one striking study, married couples admitted to a hospital research unit received small wounds on their arms, and wound healing was monitored over subsequent days. Couples who engaged in hostile arguments showed significantly slower healing than couples who maintained more supportive interactions, demonstrating how interpersonal dynamics directly affect physical health.

The quality of marital interaction appears more important than mere presence or absence of conflict. All couples have disagreements, but couples who handle conflict constructively without contempt, defensiveness, or stonewalling maintain health better than couples who engage in destructive conflict patterns. Research by psychologist John Gottman has identified specific interaction patterns that predict relationship deterioration and divorce, including criticism, contempt, defensiveness, and withdrawal. These patterns are associated not only with relationship failure but also with increased health problems.

Domestic violence, whether physical, sexual, or emotional, causes enormous suffering and health damage. Beyond immediate physical injuries, abuse victims show elevated rates of chronic pain, gastrointestinal problems, cardiovascular disease, depression, anxiety, post-traumatic stress disorder, and substance abuse. The chronic stress of living in abusive relationships appears to damage health through sustained activation of stress systems, disrupted sleep, unhealthy coping behaviors, and possibly through epigenetic changes affecting gene expression.

Children exposed to inter parental violence or abuse show similar health consequences, with effects persisting into adulthood. The adverse childhood experiences literature discussed in earlier chapters demonstrates that witnessing parental conflict or experiencing abuse creates lasting vulnerabilities to stress-related illness. These effects reflect both direct biological impacts of toxic stress during development and indirect effects through learned relationship patterns that perpetuate dysfunction across generations.

Toxic relationships extend beyond romantic partnerships to include friendships, family relationships, and workplace relationships that involve manipulation, exploitation, or chronic negativity. While less studied than marital relationships, these connections likely affect health through similar stress pathways. The concept of emotional labor, the effort required to manage emotions and present appropriate displays in relationships or work settings, captures how some relationships drain rather than sustain well being.

Social rejection, particularly when it involves exclusion, ostracism, or discrimination, activates similar neural pain circuits as physical injury and romantic heartbreak. Experimental studies using virtual ball-tossing games show that being excluded from the game activates the anterior cingulate cortex and other regions involved in physical pain even when participants know the game is trivial and the excluders are strangers. This exquisite sensitivity to social exclusion likely evolved because exclusion from social groups threatened survival in ancestral environments where individuals depended on groups for protection and resources.

Discrimination based on race, ethnicity, gender, sexual orientation, disability, or other stigmatized identities represents chronic exposure to social stress with profound health consequences. People experiencing discrimination show elevated stress hormone levels, increased inflammation, higher blood pressure, and increased rates of mental and physical illness. The biological weathering caused by chronic discrimination may contribute to health disparities, with marginalized groups showing higher rates of stress-related disease partly attributable to discrimination experiences.

Understanding the health harms of negative relationships does not negate the benefits of positive relationships but rather highlights the importance of relationship quality. The goal is not simply to avoid isolation but to cultivate supportive, nurturing relationships while setting boundaries against toxic dynamics. For people in abusive relationships, leaving may be essential for protecting health even if it temporarily increases isolation. For those in moderately conflictual relationships, improving interaction patterns through couples therapy or skills training may transform health-damaging relationships into health-promoting ones.

Technology and Social Connection in the Modern World

The rise of digital communication technology has transformed how people form and maintain social relationships, with implications for social connection and health that are still being understood. Social media platforms, text messaging, video calls, and online communities offer unprecedented opportunities to connect across distances and find like-minded others. Yet concerns persist that digital connection may substitute for rather than supplement face-to-face interaction, potentially undermining the deep relationships that promote well being.

Research on technology's effects on social connection reveals complex, mixed findings. Some studies suggest that social media use is associated with increased loneliness, depression, and anxiety, particularly among adolescents and young adults who spend extensive time on platforms. These negative effects might reflect several mechanisms. Social comparison with curated, idealized presentations of others' lives may promote feelings of inadequacy. Passive scrolling through others' content without active engagement may fail to provide the reciprocity and intimacy of real connection. Time spent on devices may displace time for in-person interaction, reducing relationship quality.

However, other research finds positive effects, particularly when technology is used to maintain existing relationships rather than substitute for them. Video calls with distant family and friends, text conversations that maintain daily contact, and online communities bringing together people with shared interests or challenges can all strengthen relationships and reduce isolation. For people with mobility limitations, social anxiety, or living in remote areas, digital connection may provide relationships that would otherwise be unavailable.

The key distinction appears to be between active, personalized communication that maintains real relationships versus passive consumption of content or superficial interactions with large networks of weak ties. A video call with a close friend likely provides similar benefits to in-person interaction, while scrolling through feeds of acquaintances' posts likely does not. Messaging back and forth with someone you care about creates connection, while broadcasting to audiences without genuine reciprocal engagement does not.

The quality rather than quantity of online social contact appears crucial, mirroring findings about offline relationships. Having hundreds or thousands of social media connections means little if those connections lack depth, trust, and reciprocity. Meanwhile, maintaining a few close relationships through regular video calls, messages, and occasional in-person visits can sustain well being even when face-to-face contact is limited.

Concerns about technology's effects on young people deserve particular attention given ongoing brain development during childhood and adolescence. Excessive screen time may displace activities important for development including physical play, in-person social interaction, sleep, and academic engagement. Social media may amplify insecurities about appearance, status, and social acceptance that are already heightened during adolescence. Cyber bullying and social exclusion can occur at scale and intensity impossible in per-digital environments.

However, technology also provides benefits for young people including access to information, creative outlets, communities of support, and connections with peers during periods like the pandemic when in-person interaction was restricted. The challenge is helping young people navigate technology in ways that support rather than undermine development and well being. This likely requires setting appropriate limits, modeling healthy technology use, maintaining open communication about online experiences, and ensuring technology supplements rather than replaces offline relationships and activities.

The neurobiological effects of digital versus in-person interaction remain incompletely understood. Some research suggests that in-person interaction triggers stronger oxytocin responses than digital communication, potentially cxplaining why video calls and messaging, while valuable, may not fully replicate the bonding effects of physical presence. The absence of physical touch, pheromones, and subtle nonverbal cues in digital communication may reduce the biological impact of interaction.

However, brain imaging studies show that text messages from romantic partners can activate reward circuits and reduce stress responses, suggesting that digital communication can engage some of the same neural systems as in-person contact. The meaning and emotional significance of communication may matter more than the medium, with heartfelt expressions of care and support providing benefits regardless of whether they occur face-to-face, by phone, or through text.

Moving forward, the challenge is integrating technology into social lives in ways that enhance rather than replace meaningful connection. This might involve using technology intentionally to maintain relationships that matter while setting boundaries against passive consumption and superficial interaction. It requires critical evaluation of how different technologies and usage patterns affect personal wellbeing rather than assuming all digital connection is equivalent. It demands recognition that while technology enables connection in powerful ways, it cannot fully replace the depth of in-person relationships built through shared physical presence over time.

Conclusion: The Biological Imperative of Connection

The evidence is overwhelming that human beings are not merely social by preference but social by biological necessity. Our brains and bodies evolved in social contexts, dependent on cooperation and connection for survival and reproduction. The neural and hormonal systems that create feelings of bonding, trust, and love are as fundamental to our biology as systems regulating hunger, thirst, and breathing. Social connection is not a luxury but a basic need, with deprivation causing suffering and health damage comparable to physical deprivation.

Understanding the biology of social connection reveals both why relationships matter so profoundly and how to cultivate the connections that promote well being. It validates the pain of loneliness and rejection while offering hope that connection can be deliberately sought and nurtured. It explains why investing in relationships is investing in health, and why social policies supporting family bonds, community cohesion, and inclusion matter not just for quality of life but for public health.

The modern world presents both challenges and opportunities for social connection. While technology, mobility, and changing social structures can fragment traditional communities and isolate individuals, they also enable new forms of connection across distances and boundaries. The task is using understanding of what humans need biologically to create lives, communities, and societies that support the deep, meaningful relationships essential for thriving.

9

HOW MEMORY WORKS

Memory is the foundation of who we are. Without memory, we would have no sense of personal history, no recognition of friends and family, no knowledge of language, and no ability to learn from experience. We would live in an eternal present, each moment disconnected from the past and unable to inform the future. The profound importance of memory becomes heartbreakingly clear in diseases like Alzheimer's, where progressive memory loss gradually strips away identity, relationships, and independence, leaving patients unable to recognize loved ones or remember events from moments earlier.

Yet memory is far more complex and far stranger than most people realize. It is not a video recording that faithfully preserves experiences for later playback. Rather, memory is a reconstructive process that creates representations of the past from fragments of information, filling gaps with inferences and interpretations that can transform memories in subtle or dramatic ways. The confidence we feel in our memories often exceeds their accuracy. Vivid, detailed memories can be substantially or entirely false. Different memory systems operate according to different principles, with some memories becoming more accessible with time while others fade. Understanding how memory actually works reveals both its remarkable capabilities and its surprising limitations.

The scientific study of memory was revolutionized by the case of Henry Molaison, mentioned briefly in Chapter Two. In 1953, at age twenty-seven, Molaison underwent experimental brain surgery to treat severe epilepsy. Surgeons removed large portions of his medial temporal lobes, including most of the hippocampus, on both sides of his brain. The surgery successfully reduced his seizures, but it produced devastating and unexpected consequences for his memory that would make him one of the most studied patients in the history of neuroscience.

After the surgery, Molaison could no longer form new long-term memories. He could hold information in mind briefly through active rehearsal, maintaining a conversation or remembering a number for a few minutes. But as soon as his attention shifted, the information vanished completely. He could not remember what he had eaten for breakfast, could not recognize his doctors despite meeting them hundreds of times, could not learn that his parents had died. He remained frozen in time, with his most recent intact memories from several years before the surgery. Each day, he woke experiencing the same shock of seeing his aged reflection in the mirror, unable to remember the gradual changes of intervening decades.

Yet Molaison's memory was not entirely destroyed. His memories from before the surgery remained largely intact, demonstrating that the hippocampus is not where long-term memories are permanently stored. He could also learn new motor skills like drawing in a mirror or tracing shapes, improving with practice even though he had no conscious memory of practicing. This revealed that not all forms of memory depend on the hippocampus, and that the brain contains multiple memory systems operating independently.

This chapter explores how memory works at multiple levels, from the molecular and cellular mechanisms that create lasting changes in neurons, to the brain systems that encode, consolidate, and retrieve different types of memories, to the psychological processes that make memory both remarkably capable and surprisingly fallible. We will examine why some experiences are vividly remembered while others are forgotten, how memories change over time, why memory sometimes fails, and what this understanding reveals about learning, identity, and the nature of human experience.

Types of Memory: A System of Systems

One of the most important insights from memory research is that memory is not a single unified system but rather a collection of different systems that operate according to different principles and depend on different brain structures. Understanding these distinctions is crucial for understanding both how memory works normally and what goes wrong in various forms of memory impairment.

The broadest distinction separates short-term memory from long-term memory. Short-term memory, also called working memory in its more sophisticated forms, holds information temporarily for immediate use. When you look up a phone number and keep it in mind long enough to dial, you are using short-term memory. When you listen to the beginning of a sentence and hold those words in memory while waiting to hear how the sentence ends, you are using working memory. These temporary memory systems have limited capacity, typically holding only about seven items or chunks of information at a time, and limited duration, with information fading within seconds unless actively maintained through rehearsal.

Working memory is not merely a passive storage system but an active workspace where information can be maintained and manipulated. When you solve mental arithmetic problems, follow directions to a new location, or keep track of conversational topics, you engage working memory systems. The prefrontal cortex plays crucial roles in working memory, maintaining representations of information even when it is no longer perceptually present and manipulating that information to accomplish goals.

Long-term memory, in contrast, has essentially unlimited capacity and can store information for duration ranging from minutes to entire lifetimes. The distinction between short-term and long-term memory is not merely about duration but reflects fundamentally different memory systems. Information in short-term memory exists as active neural firing patterns that cease when attention shifts. Information in long-term memory requires creation of lasting changes in neural connections through processes of consolidation that can take hours, days, or even longer.

Within long-term memory, the most important distinction separates declarative memory from non-declarative memory. Declarative memory, also called explicit memory, is memory for facts and events that can be consciously recalled and verbally declared. This is the type of memory most people think of when they use the word memory: remembering what you did yesterday, recalling historical facts, recognizing people you know. Declarative memory can be further divided into episodic memory for personal experiences located in time and place, and semantic memory for general knowledge and facts not tied to specific learning episodes.

Episodic memory allows you to mentally travel back in time to re-experience events from your past. When you remember your last birthday party, your first day of school, or what you had for dinner last night, you are accessing episodic memories. These memories include not just what happened but also contextual details about when and where events occurred, who was present, how you felt, and other features that create a sense of mentally reliving the experience. Episodic memory is what gives you a sense of personal history and continuous identity across time.

Semantic memory contains general knowledge about the world that is not tied to remembering specific learning episodes. You know that Paris is the capital of France, that triangles have three sides, that dogs are mammals, but you probably cannot remember the specific occasions when you learned these facts. Semantic knowledge accumulates gradually through repeated exposure and becomes decontextualized, losing connections to the times and places where learning occurred. While episodic and semantic memory are theoretically distinct, they often work together, with episodic memories contributing to the gradual accumulation of semantic knowledge.

The hippocampus and surrounding medial temporal lobe structures are essential for forming new declarative memories, as dramatically demonstrated by Henry Molaison's case. When you experience an event or learn a fact, the hippocampus rapidly creates associations among the distributed cortical regions that represent different aspects of the experience. A memory of having coffee with a friend involves visual representations of your friend's face, auditory representations of their voice, spatial representations of where you met, emotional representations of how you felt, semantic representations of what you discussed. The hippocampus binds these distributed elements together, creating a coherent memory that can later be retrieved.

Non-declarative memory, also called implicit memory, includes several forms of memory that influence behavior without requiring conscious recollection. These memory systems operate outside awareness and cannot be verbally described. The most extensively studied form is procedural memory for skills and habits. When you learn to ride a bicycle, touch-type, play a musical instrument, or drive a car, you are forming procedural memories. These skills improve with practice and become automatic, no longer requiring conscious attention or thought.

Procedural memory depends on brain structures including the basal ganglia and cerebellum rather than the hippocampus. This is why Henry Molaison could learn new motor skills even though he could not form new declarative memories. The independence of procedural and declarative memory systems explains why skilled performance often becomes harder when you try to consciously think about what you are doing. The procedural system works best when left to operate automatically without interference from conscious deliberation.

Other forms of non-declarative memory include priming, where prior exposure to stimuli influences responses to those stimuli even without conscious memory of the exposure, and classical conditioning, where associations are learned between stimuli or between stimuli and responses. These memory systems involve various brain structures depending on the specific type of learning, but they share the property of affecting behavior without requiring conscious recollection.

Understanding these multiple memory systems explains many puzzling phenomena. Patients with hippocampal damage like Molaison lose the ability to form new declarative memories but retain procedural learning. Conversely, patients with basal ganglia damage from Parkinson's disease or Huntington's disease show impaired procedural learning but intact declarative memory. People can accurately perform tasks they have no conscious memory of practicing. Preferences and emotional responses can be influenced by experiences that are not consciously remembered. Memory is not one thing but many things, each serving different functions and following different rules.

Encoding: How Experiences Become Memories

For an experience to be remembered later, it must first be encoded, transformed from a fleeting perceptual experience into a more lasting neural representation. Encoding is not a passive recording process but an active construction that depends on attention, elaboration, and integration with existing knowledge. Understanding the factors that promote effective encoding reveals why some experiences are easily remembered while others are quickly forgotten.

Attention is the first requirement for effective encoding. Information that does not receive attention is unlikely to be encoded into long-term memory.

This explains the common experience of forgetting where you placed your keys or what someone just said: if you were not paying attention when you set down the keys or heard the words, no memory was created to later retrieve. The limits of attention mean that we cannot encode everything we experience, and what gets encoded depends on where attention is directed.

The phenomenon of inattentional blindness demonstrates how completely unattended information can fail to be encoded. In famous demonstrations, people watching a video of a basketball game and counting passes often fail to notice a person in a gorilla suit walking through the scene. The gorilla is clearly visible and spends several seconds on screen, yet roughly half of observers completely miss it because their attention is focused on counting passes. This shows that encoding depends not on whether information is perceived but on whether it receives attention.

Depth of processing powerfully influences encoding success. Information that is processed superficially, focusing only on physical characteristics like visual appearance or sound, is encoded weakly and forgotten quickly. Information that is processed deeply, focusing on meaning, significance, and connections to existing knowledge, is encoded strongly and remembered well. When you read a word and simply notice its font or length, shallow processing creates weak memory. When you think about what the word means, how it relates to your life, or how it connects to things you already know, deep processing creates strong memory.

Elaboration, the process of enriching encoding through associations, imagery, and connections, substantially improves memory. When you try to remember a person's name, simply repeating it creates minimal encoding. But if you notice that their name rhymes with something, reminds you of someone else you know, or can be visualized in some meaningful way, you create multiple retrieval routes that make the memory more accessible later. Memory improvement techniques like the method of loci, where items to be remembered are mentally placed in familiar locations, work by forcing elaborative encoding.

Organization also facilitates encoding and retrieval. Information that is structured, categorized, or organized according to meaningful schemes is remembered better than random, disorganized information. This is why students who organize course material into coherent frameworks and outlines typically learn better than those who try to memorize disconnected facts. The act of organizing information requires processing relationships and hierarchies, creating a structured memory representation that supports efficient retrieval.

Emotional arousal powerfully enhances memory encoding, as discussed in earlier chapters on emotion. Emotionally significant events are typically remembered better than emotionally neutral events. This effect is mediated by the amygdala, which modulates memory consolidation in the hippocampus and other brain regions. When the amygdala is activated by emotional arousal, it enhances memory formation through release of stress hormones and neurotransmitters that strengthen encoding.

However, the relationship between emotion and memory is complex. While emotional arousal enhances memory for the central features of events, it can actually impair memory for peripheral details. This creates a paradox where traumatic memories may be vivid for core elements of what happened but have gaps for surrounding context. Additionally, extreme stress can impair hippocampal function, potentially fragmenting memories or impairing their organization. This may contribute to the disorganized, fragmented quality of traumatic memories in post-traumatic stress disorder.

Distinctiveness, the extent to which information stands out from other information, promotes strong encoding. Events that are unusual, surprising, or different from routine experiences are more likely to be remembered. This is sometimes called the Von Restorff effect after the researcher who first documented it. In a list of similar items, a distinctive item that differs in some way is remembered better than the others. This principle explains why novel experiences, unexpected events, and first occurrences tend to be particularly memorable.

The serial position effect describes the finding that items at the beginning and end of lists or sequences are remembered better than items in the middle. Items early in a sequence benefit from primacy effects, with more rehearsal and deeper processing. Items late in a sequence benefit from recency effects, with information still in short-term memory at the time of testing. Items in the middle receive less rehearsal than early items and have more subsequent items interfering with their retention, making them most vulnerable to forgetting.

Generation effects demonstrate that information we generate ourselves is remembered better than information we passively receive. When you create your own examples, generate your own questions, or produce your own summaries, you remember information better than when you simply read or hear it. This active generation requires deeper processing and creates more retrieval routes. The principle has important implications for education, suggesting that learning environments encouraging active production of knowledge create stronger memories than passive reception of information.

Context during encoding powerfully influences later memory. The environmental, emotional, and internal state context in which learning occurs becomes associated with the learned information. This creates context-dependent memory, where memory is best when retrieval occurs in the same context as encoding. Divers who learned word lists underwater recalled them better when tested underwater than on land, while divers who learned on land recalled better when tested on land. Similarly, material learned in a particular room, mood state, or physiological condition is often best recalled in that same context.

These encoding principles have practical implications for improving memory. To remember something well, pay full attention when trying to learn it. Process information deeply, thinking about meaning rather than superficial characteristics. Elaborate through associations, imagery, and connections to existing knowledge. Organize information into meaningful structures. Engage emotionally with material when possible. Make information distinctive. Generate your own examples and explanations rather than passively receiving information. Study in conditions similar to those where memory will be tested.

Consolidation: Stabilizing Memories Over Time

Encoding creates an initial memory trace, but this trace is fragile and vulnerable to disruption. The process of consolidation gradually stabilizes memories, making them more resistant to interference and forgetting. Consolidation occurs at multiple timescales and involves transfer of information from temporary to more permanent storage, with sleep playing particularly important roles.

At the cellular level, consolidation involves structural changes in synaptic connections between neurons. As discussed in Chapter Three, learning strengthens some synaptic connections through long-term potentiation while weakening others through long-term depression. These changes initially involve modifications to existing proteins at synapses, but lasting memory requires synthesis of new proteins that produce more permanent structural changes including growth of new synaptic connections and dendritic spines.

This protein synthesis-dependent consolidation takes several hours to complete, creating a window during which new memories remain vulnerable. Disrupting protein synthesis during this consolidation window prevents formation of long-term memories even though short-term memory remains intact. Head injuries that cause temporary loss of consciousness often produce retrograde amnesia for events occurring shortly before the injury, presumably because consolidation had not completed when brain function was disrupted.

At the systems level, consolidation involves gradual transfer of memories from the hippocampus to the neocortex. The hippocampus rapidly encodes new memories, binding together distributed cortical representations as described earlier. But with time and repeated retrieval, memories become less dependent on the hippocampus and more dependent on direct connections among cortical regions that represent memory content. This systems consolidation process can take months or years to complete.

Evidence for systems consolidation comes from studies of patients with hippocampal damage. Such patients typically show temporally graded amnesia, with recent memories more impaired than remote memories. Henry Molaison could remember his childhood and young adulthood reasonably well but had no memories from the years immediately before his surgery and could not form new memories afterward. This pattern suggests that old memories had undergone systems consolidation and no longer depended on the hippocampus, while recent memories still required hippocampal involvement.

Sleep plays crucial roles in memory consolidation, with numerous studies showing that sleep deprivation after learning impairs later memory. During sleep, particularly during slow-wave sleep and REM sleep, the brain appears to replay recent experiences, reactivating the neural patterns that were active during initial learning. This reactivation seems to drive consolidation, strengthening synaptic changes and promoting transfer of information to cortical storage.

Brain imaging and electrophysiological studies have captured this replay process. Neurons in the hippocampus that fired in particular sequences while rats navigated a maze fire in the same sequences during subsequent sleep, but at much faster speeds as if the brain is reviewing the day's experiences in fast-forward. Similar replay occurs in humans, with brain regions active during learning showing reactivation during sleep. The amount of replay predicts how well information will be remembered, suggesting this process is functionally important for consolidation.

The benefits of sleep for memory consolidation have practical implications. Students who study and then sleep before being tested typically remember information better than students who study and remain awake for the same duration before testing. Sleeping soon after learning appears more beneficial than delayed sleep. Naps can enhance memory consolidation, with even brief periods of sleep providing benefits. These findings challenge the common practice of cramming before exams without sleep, as sacrificing sleep to study may actually impair memory for what is studied.

Different stages of sleep may contribute to consolidating different types of memories. Slow-wave sleep, the deep sleep characterized by large, slow electrical waves in the brain, appears particularly important for consolidating declarative memories. REM sleep, the stage characterized by rapid eye movements, vivid dreams, and brain activity resembling waking, may preferentially consolidate procedural and emotional memories. Getting sufficient amounts of both slow-wave and REM sleep may be necessary for optimal memory consolidation.

Emotional memories appear to undergo preferential consolidation, becoming even more prominent in memory over time. Sleep seems to selectively consolidate emotional aspects of memories while allowing neutral details to fade. This creates a phenomenon where emotional memories can become increasingly isolated from their original contexts, potentially contributing to anxiety disorders and PTSD where emotional reactions persist even when conscious memory of triggering events is incomplete.

Stress and cortisol can both enhance and impair consolidation depending on timing and intensity. Moderate stress and cortisol elevation shortly after learning can enhance consolidation, potentially explaining why moderately stressful or emotionally arousing events are well remembered. However, extreme stress or cortisol elevation can impair hippocampal function and disrupt consolidation, potentially creating fragmented memories. The inverted-U relationship between arousal and memory consolidation helps explain why moderate emotional arousal enhances memory while extreme stress can impair it.

Retrieval: Accessing Stored Memories

Even perfectly encoded and consolidated memories are useless if they cannot be retrieved when needed. Retrieval, the process of accessing stored memories and bringing them to consciousness, is itself an active process that can strengthen, modify, or distort memories. Understanding retrieval reveals both why memory sometimes fails despite information being stored, and how the act of remembering can change what is remembered.

Retrieval depends heavily on cues, environmental or internal signals that activate associated memories. When you smell a particular perfume and suddenly remember a person who wore it, when hearing a song transports you back to a specific time in your life, or when visiting a childhood home triggers floods of forgotten memories, you are experiencing cue-dependent retrieval. The more overlap between cues present during encoding and cues available during retrieval, the more likely memory will be successfully accessed.

The encoding specificity principle, proposed by Endel Tulving, states that memory is best when cues available during retrieval match those present during encoding. This explains context-dependent memory effects discussed earlier, where returning to the learning environment facilitates retrieval. It also explains state-dependent memory, where internal physiological or emotional states serve as cues. Information learned while in a particular mood or under the influence of substances like alcohol or caffeine is sometimes better recalled when in the same state, though substance-induced impairment generally harms memory overall.

Retrieval mode refers to the cognitive state of trying to remember, which involves activating brain networks including the prefrontal cortex and hippocampus that search for relevant memories. Simply being in retrieval mode, actively trying to remember, is sometimes sufficient to begin accessing memories even without specific cues. However, retrieval mode can also produce false memories when searches retrieve information that seems relevant but is actually incorrect or from different episodes.

Recognition and recall represent different forms of retrieval requiring different amounts of information. Recognition involves judging whether something was previously encountered, typically easier than recall because the test item itself serves as a powerful retrieval cue. When you see someone at a party and recognize them as familiar but cannot remember their name or where you met them, you are experiencing successful recognition without successful recall. Recall requires generating information from memory without the item being present, typically harder because you must generate effective retrieval cues yourself.

The testing effect, also called retrieval practice, demonstrates that actively retrieving information strengthens memory more than simply studying or re-reading material. When students practice retrieving information through self-testing, they later remember information better than students who spent the same amount of time re-studying. This effect occurs because retrieval itself is a learning event that strengthens memory traces and creates additional retrieval routes.

However, retrieval is not simply reactivation of stored representations but a reconstructive process that can modify memories. Each time a memory is retrieved, it enters a labile state during a process called re consolidation, where it can be strengthened, updated with new information, or modified before being restored to long-term storage. This re consolidation process allows memories to be updated with current information but also creates opportunities for memories to be distorted or contaminated with information from other sources.

Source monitoring errors occur when people correctly remember information but incorrectly remember where the information came from. You might remember a fact but forget whether you experienced it firsthand, heard it from someone else, or imagined it. You might remember an event but confuse whether it happened to you or to someone you heard about. These source errors demonstrate that memory stores content and source information separately, and source information is particularly vulnerable to forgetting or confusion.

False memories, confidently held memories for events that never occurred, demonstrate the reconstructive nature of memory. Extensive research by Elizabeth Loftus and others has shown how easily false memories can be created through suggestive questioning, imagination, or exposure to misleading information. In the misinformation effect, exposure to misleading information after an event can cause people to misremember what they actually experienced. Witnesses who hear other witnesses describe an event sometimes incorporate those descriptions into their own memories, confidently remembering details they never actually perceived.

More dramatically, entirely false memories for childhood events can sometimes be implanted through repeated suggestions and imagination exercises. In one study, about a quarter of participants developed false memories of being lost in a shopping mall as a child after family members suggested this had happened and participants were encouraged to imagine and elaborate the event. These false memories often included rich sensory details and were held with high confidence, indistinguishable from true memories in subjective experience though they never occurred.

The malleability of memory has important implications, particularly for eyewitness testimony in legal proceedings. Eyewitnesses often testify with great confidence about events they claim to remember clearly, and juries tend to believe confident eyewitnesses. However, research shows that confidence and accuracy are only weakly related, and confident witnesses can be entirely mistaken. Suggestive police interviews, exposure to media coverage, conversations with other witnesses, and the mere passage of time can all contaminate eyewitness memories without witnesses being aware their memories have changed.

Understanding these limitations has led to reforms in how police conduct lineups and interviews to minimize suggestion and preserve memory accuracy. Sequential lineups where suspects are presented one at a time produce fewer false identifications than simultaneous lineups where all options are visible at once. Cognitive interviews that use open-ended questions and avoid leading suggestions produce more accurate information than traditional interrogative interviews. Recording of interviews allows later evaluation of whether procedures might have contaminated memory.

The reconstructive nature of memory does not mean memory is useless or that we should distrust all memories. Most memories are reasonably accurate for core details even if peripheral details are forgotten or misremembered. The gist of experiences is typically preserved better than exact details. Memories for highly distinctive, personally significant events are generally more reliable than memories for routine events. But memory should not be treated as infallible, and confidence should not be equated with accuracy.

Forgetting: When and Why Memory Fails

Forgetting is often viewed as a failure of memory, an unfortunate limitation to be overcome. However, forgetting serves important adaptive functions. If we remembered everything with perfect detail, our minds would be cluttered with vast amounts of irrelevant information that would interfere with finding relevant memories. Forgetting allows us to discard outdated information, generalize across specific episodes to extract common patterns, and move beyond painful or no-longer-relevant experiences. Understanding the mechanisms of forgetting reveals that memory loss is not simply passive decay but often reflects active processes that shape what is retained.

The forgetting curve, first described by Hermann Ebbinghaus in the nineteenth century, shows that forgetting is most rapid soon after learning and then gradually slows. Much of what we learn is forgotten within hours or days, but information that survives this initial rapid forgetting tends to be retained for longer periods. This pattern makes adaptive sense: recently encountered information is most likely to be relevant soon, so rapid forgetting quickly discards irrelevant information while preserving information that proves useful enough to be retrieved repeatedly.

Decay theories propose that memories fade with time if they are not used, similar to how footprints in sand gradually disappear. While some forgetting likely involves passive decay of synaptic changes, decay alone cannot explain forgetting patterns. Information that is not retrieved for long periods can sometimes be suddenly recalled when appropriate cues are encountered, suggesting it was stored but inaccessible rather than truly lost. Additionally, the passage of time per se seems less important than what happens during that time.

Interference theories propose that forgetting occurs because new learning interferes with old memories, or old memories interfere with new learning. Proactive interference occurs when previously learned information makes it harder to learn and remember new information. Retroactive interference occurs when newly learned information makes it harder to remember previously learned information. Both forms of interference are well-documented in laboratory studies and likely contribute substantially to everyday forgetting.

Retrieval-induced forgetting demonstrates that the very act of retrieving some memories can cause forgetting of related but non-retrieved memories. When you repeatedly retrieve certain memories from a category while ignoring others, the retrieved memories become stronger while the ignored memories become weaker. This might occur because retrieval activates inhibitory processes that suppress competing memories. While this seems like a design flaw, it may actually be adaptive, allowing frequently retrieved memories to become more accessible while suppressing less relevant competing memories.

Motivated forgetting occurs when people intentionally or unconsciously avoid retrieving unwanted memories. The concept of repression, central to psychoanalytic theory, proposes that traumatic or threatening memories are unconsciously banished from awareness. While evidence for unconscious repression remains controversial, research clearly shows that people can intentionally suppress unwanted memories through effortful inhibition. Repeatedly suppressing retrieval of specific memories makes those memories less accessible later, suggesting intentional forgetting is possible though not always successful.

The neural mechanisms of intentional forgetting involve prefrontal regions inhibiting hippocampal retrieval processes. Brain imaging studies show that when people try to suppress memory retrieval, prefrontal regions become active while hippocampal activity decreases. With practice, suppression becomes more effective and requires less prefrontal effort, suggesting the brain learns to automatically inhibit unwanted memories.

Infantile amnesia, the inability to remember experiences from early childhood, particularly before age three or four, has puzzled psychologists for over a century. Several factors contribute to this early memory gap. The hippocampus and prefrontal cortex continue developing throughout childhood, with immature systems less capable of forming and retrieving memories. Young children lack the language and narrative skills needed to organize experiences into lasting episodic memories. The sense of self required for autobiographical memory develops gradually. Rapid brain growth and learning during early childhood may involve active forgetting that allows neural circuits to be reorganized.

However, infantile amnesia does not mean early experiences leave no trace.

Procedural memories, emotional conditioning, and attachment patterns all form during infancy and influence later development despite lack of conscious recollection. The effects of early trauma persist even when explicit memories are unavailable. This dissociation between lack of conscious memory and presence of lasting effects demonstrates again that multiple memory systems operate independently.

Pathological forgetting occurs in various conditions including amnesia, dementia, and traumatic brain injury. Amnesia refers to severe memory impairment caused by brain damage or psychological trauma. Retrograde amnesia is loss of memory for events before the brain injury or onset of illness. Anterograde amnesia is inability to form new memories after onset. The extent and pattern of amnesia depends on what brain structures are damaged and how severely.

Alzheimer's disease, the most common cause of dementia, involves progressive neurodegeneration that destroys memory. Early in the disease, episodic memory for recent events is impaired while remote memories and procedural memories remain relatively preserved. As the disease progresses, semantic memory deteriorates, language abilities decline, and eventually even remote memories are lost. The progression reflects the spread of pathology from hippocampus and entorhinal cortex to wider cortical regions.

The devastation of Alzheimer's disease highlights how central memory is to identity and function. When memory fails, people lose not just knowledge but their sense of self, their relationships, and their independence. Understanding the biological basis of memory in health provides hope for developing treatments that might prevent or slow memory loss in disease.

Memory Across the Lifespan

Memory abilities change substantially across the lifespan, from the rapid learning of childhood through the memory changes of normal aging and the pathological memory loss of neurodegenerative disease. Understanding these changes reveals how brain development and aging affect cognition and what can be done to maintain cognitive health throughout life.

Young children show remarkable memory abilities in some domains while showing limitations in others. Infants can recognize faces, voices, and objects after brief exposures, demonstrating impressive perceptual memory. They rapidly learn language, procedural skills, and emotional associations. However, their episodic memory is limited, with poor memory for specific events and contexts. Semantic memory for facts and concepts develops gradually as language and conceptual understanding mature.

Memory continues improving throughout childhood and adolescence as the hippocampus and prefrontal cortex develop. Children become better at using memory strategies including rehearsal, organization, and elaboration. They develop metamemory, understanding of their own memory processes that allows them to judge what they know and allocate study time effectively. Working memory capacity increases, supporting more complex reasoning and learning.

Young adulthood represents the peak of memory performance for most types of memory. Processing speed, working memory capacity, and episodic memory are all maximal in the twenties and early thirties. However, even young adults show forgetting patterns similar to those described by Ebbinghaus, and even optimal memory systems are far from perfect recorders of experience.

Normal aging brings gradual changes in memory that begin in the forties or fifties and continue throughout later life. Episodic memory typically shows the most decline, with older adults having particular difficulty remembering where and when events occurred. Working memory capacity decreases, making it harder to hold and manipulate multiple pieces of information simultaneously. Processing speed slows, affecting how quickly information can be encoded and retrieved.

However, aging does not affect all aspects of memory equally. Semantic memory often remains stable or even improves with age as accumulated knowledge grows. Procedural memory for well-learned skills is largely preserved. Recognition memory holds up better than recall. Memory for gist and meaning is preserved better than memory for specific details. These patterns suggest that aging affects particular memory processes and brain systems while leaving others relatively intact.

The neural basis of age-related memory decline involves multiple factors. The hippocampus shows structural changes including volume loss and reduced neurogenesis. Prefrontal cortex structure and function change, reducing working memory capacity and strategic processing. White matter tracts connecting brain regions degenerate, slowing information transfer. Neurotransmitter systems including acetylcholine and dopamine show reduced function. Accumulation of pathology including amyloid plaques and tau tangles, even in cognitively normal elders, may contribute to subtle memory changes.

However, age-related memory decline is highly variable, with some individuals showing minimal change while others show substantial impairment. Factors protecting cognitive function in aging include education, mentally stimulating activities, physical exercise, social engagement, good cardiovascular health, and absence of depression. These protective factors suggest that cognitive aging is not rigidly determined but can be influenced by lifestyle and health behaviors.

The distinction between normal cognitive aging and pathological conditions like Alzheimer's disease can be challenging but is crucial. Normal aging involves gradual, relatively mild memory changes that do not substantially interfere with independence or daily functioning. Pathological conditions involve more severe, progressive decline that impairs ability to work, manage finances, take medications, or live independently. Memory complaints in normal aging typically involve difficulty retrieving names or details but eventual retrieval with effort or cues. In Alzheimer's disease, information is truly lost, with people unable to remember even with extensive cues and often unaware they have forgotten.

Efforts to maintain memory and cognitive function throughout the lifespan focus on modifiable factors. Physical exercise, particularly aerobic exercise, promotes brain health through multiple mechanisms including increased blood flow, enhanced neurogenesis, and release of growth factors. Mental stimulation through learning, puzzles, reading, and other cognitively demanding activities may build cognitive reserve that helps compensate for age-related brain changes. Social engagement provides cognitive stimulation and stress buffering. Management of cardiovascular risk factors including hypertension, diabetes, and high cholesterol protects brain health. Adequate sleep supports memory consolidation. Mediterranean-style diets rich in fruits, vegetables, whole grains, and healthy fats are associated with better cognitive aging.

While no interventions have been conclusively proven to prevent Alzheimer's disease, evidence suggests that combinations of these healthy lifestyle factors reduce risk. The concept of cognitive reserve proposes that education, mentally stimulating careers, and lifelong learning build neural networks and processing efficiency that allow the brain to maintain function despite accumulating pathology. This provides hope that people can influence their cognitive trajectories through choices and behaviors throughout life.

Conclusion: Memory as Construction and Reconstruction

Memory is simultaneously one of our most impressive cognitive abilities and one of our most fallible. We can remember detailed information from decades past, learn vast amounts of new information, and use memory to navigate complex environments and relationships. Yet we also forget most of what we experience, misremember details of important events, and sometimes develop confident memories for events that never occurred.

Understanding memory as a reconstructive process rather than a recording device helps make sense of these apparent contradictions. Memory evolved not to create perfect records of the past but to extract and preserve information useful for navigating the future. This requires balancing preservation of important information against forgetting of irrelevant details, updating memories with new information while maintaining some stability, and constructing coherent narratives from fragmented information even if this sometimes introduces errors.

The biological understanding of memory reveals both the capabilities and limitations of memory systems. It explains why some experiences are vividly remembered while others are quickly forgotten, why memory improves with certain practices and deteriorates with others, why different types of brain damage produce different patterns of memory impairment, and why memory naturally changes across the lifespan. This knowledge empowers people to use memory more effectively while recognizing its limitations and vulnerabilities.

10

LEARNING AND BRAIN PLASTICITY

The human brain contains approximately 86 billion neurons, each forming thousands of connections with other neurons, creating a network of staggering complexity. For most of the twentieth century, scientists believed that this network was essentially fixed in adulthood, with the basic structure of the brain determined during development and remaining largely unchanged thereafter. This view held that we are born with all the neurons we will ever have, that neural connections are established during critical periods in early development, and that adult brains are static structures incapable of significant reorganization.

This belief has been dramatically overturned. We now know that the brain remains plastic throughout life, constantly reorganizing its structure and function in response to experience. New synaptic connections form, existing connections strengthen or weaken, entire brain regions can reorganize to take on new functions, and under certain circumstances, new neurons can even be born in adult brains. This neuroplasticity, the brain's ability to change itself, is not a minor feature but a fundamental property that underlies all learning, enables recovery from injury, and allows us to adapt to changing environments throughout the lifespan.

The discovery of adult neuroplasticity required abandoning deeply held assumptions about brain organization. It required new technologies that could track changes in living brains over time. It required careful observations of how brains reorganize after injury or in response to intensive training. Most fundamentally, it required recognizing that the brain is not a computer with fixed hardware running different software, but rather a biological organ that physically transforms itself based on how it is used.

The implications of neuroplasticity extend far beyond neuroscience. Understanding that brains change based on experience has transformed education, suggesting that intelligence is not fixed but can be developed through effort and appropriate challenge. It has revolutionized rehabilitation, showing that functions lost to stroke or injury can sometimes be recovered through intensive retraining that recruits surviving brain regions. It has influenced how we think about expertise, revealing that exceptional abilities reflect years of practice that literally reshape the brain rather than simply inborn talent. It has even raised philosophical questions about personal identity, suggesting that we are not fixed entities but constantly changing beings shaped by our experiences.

This chapter explores the mechanisms and manifestations of neuroplasticity, examining how learning changes the brain at multiple scales from molecules to systems, what factors promote or limit plasticity, how plasticity changes across the lifespan, and how understanding plasticity can be harnessed to enhance learning and promote recovery from injury.

Mechanisms of Plasticity: How Experience Changes the Brain

At the most fundamental level, learning involves changes in the strength of synaptic connections between neurons. When two neurons are repeatedly active together, the synapse connecting them tends to strengthen, making future communication between those neurons more effective. This principle, often summarized as "neurons that fire together wire together," was first proposed by psychologist Donald Hebb in 1949 and is formalized in Hebbian learning theory.

The cellular mechanism underlying Hebbian learning is long-term potentiation, abbreviated LTP, introduced briefly in Chapter Nine. LTP occurs when repeated stimulation of a synapse causes lasting increases in the synapse's strength. This strengthening involves multiple changes. Initially, existing receptors on the postsynaptic neuron become more responsive, allowing stronger signals to pass across the synapse. With continued activity, additional receptors are inserted into the postsynaptic membrane, creating more sites for neurotransmitter binding. The presynaptic terminal may release more neurotransmitter. With sustained activity over hours and days, new dendritic spines grow, creating additional synaptic connections between the neurons.

These structural changes require synthesis of new proteins, as discussed in the previous chapter on memory consolidation. Genes in the neuron's nucleus must be activated to produce messenger RNAs that are transported to synapses where they direct protein synthesis. These proteins include structural components that enlarge and stabilize synapses, signaling molecules that facilitate neurotransmission, and receptors that make the neuron more responsive to inputs. The requirement for protein synthesis creates a molecular distinction between short-term changes that last minutes to hours and long-term changes that can last days, months, or even a lifetime.

The opposite process, long-term depression or LTD, occurs when synapses are weakly or asynchronously activated, causing lasting decreases in synaptic strength. LTD involves removal of receptors from the postsynaptic membrane, retraction of dendritic spines, and ultimately elimination of synaptic connections. This weakening and pruning of unused synapses is as important as strengthening of active synapses, allowing neural circuits to become more refined and efficient by eliminating unnecessary connections.

The balance between LTP and LTD allows neural circuits to be sculpted by experience. Frequently used pathways strengthen while rarely used pathways weaken and disappear. This creates neural networks optimized for the specific tasks and environments an individual encounters. A violinist develops enhanced representations of the left fingers that press strings, a London taxi driver develops enlarged hippocampal regions supporting spatial navigation, and a person learning a second language develops enhanced neural responses to sounds in that language. Each person's brain becomes uniquely shaped by their particular experiences.

Beyond changes in synaptic strength, neuroplasticity involves reorganization at larger scales. Cortical maps, the representations of sensory surfaces and body parts in the brain, can reorganize substantially based on experience. The somatosensory cortex contains a topographic map of the body surface, with adjacent body parts represented in adjacent cortical regions, though the map is distorted with disproportionately large representations of sensitive areas like fingers and lips. This map is not fixed but changes based on how the body is used.

Studies with monkeys demonstrate dramatic cortical reorganization following changes in sensory input. When a monkey's finger is amputated, the cortical region that previously represented that finger does not remain silent but is gradually taken over by representations of neighboring fingers. Conversely, when a monkey repeatedly uses particular fingers in a trained task requiring fine tactile discrimination, the cortical representations of those fingers expand, taking over territory from less-used regions. These changes occur over days to weeks, demonstrating that adult cortical organization is dynamic rather than static.

In humans, similar reorganization occurs following amputation, with the cortical region previously representing the lost limb being invaded by representations of adjacent body parts. This reorganization may contribute to phantom limb sensations, where amputees continue to feel the missing limb.

Because face representations in somatosensory cortex are adjacent to hand representations, hand amputation often results in face regions expanding into former hand territory. This creates the bizarre phenomenon where touching the face can produce sensations referred to the phantom hand, as signals from the face activate neurons that the brain still interprets as representing the hand.

Musicians provide dramatic examples of experience-dependent cortical reorganization. String instrument players show enlarged cortical representations of the left hand fingers used to press strings, with the magnitude of expansion correlating with the age at which training began and the amount of practice. Keyboard players show enlarged representations of both hands. Drummers show enhanced motor representations and cerebellar changes supporting precise rhythmic timing. These changes demonstrate that intensive, prolonged practice literally reshapes brain structure.

The visual system also shows remarkable plasticity. People who are blind from early in life show recruitment of visual cortex for processing auditory and tactile information. The occipital cortex, which normally processes vision, becomes activated during Braille reading, auditory localization, and other tasks in blind individuals. This cross-modal plasticity demonstrates that cortical regions are not rigidly dedicated to specific senses but can process whatever information they receive, with their function determined by their inputs rather than by predetermined specialization.

Language learning provides another domain where plasticity is clearly evident. Young children learn languages effortlessly, developing native-like pronunciation and grammar with minimal explicit instruction. Adults can certainly learn new languages but typically struggle more, often retaining accents and making grammatical errors that native speakers rarely make. This difference reflects age-related changes in plasticity, with language learning recruiting different neural systems depending on age of acquisition.

Bilingual individuals who learned both languages early in childhood show overlapping neural representations for both languages, with similar brain regions activated regardless of which language is being used. Late bilinguals who learned a second language in adulthood show more distinct neural representations, with additional brain regions recruited for the later-learned language. These differences suggest that early language learning creates more efficient, integrated neural representations than later learning, though both eventually produce functional language abilities.

Critical Periods: When Plasticity is Greatest

While the brain remains plastic throughout life, plasticity is not constant but varies across development and across brain systems. Critical periods or sensitive periods are windows of time during development when neural circuits are especially plastic and shaped by experience. During critical periods, experiences have particularly powerful and lasting effects on brain organization, while outside critical periods, the same experiences may have weaker or negligible effects.

The visual system provides the clearest examples of critical periods, largely due to pioneering work by David Hubel and Torsten Wiesel, who won the Nobel Prize for their discoveries about visual cortex development. Hubel and Wiesel studied how visual experience shapes development of neurons in the primary visual cortex that respond to inputs from each eye. They found that if one eye is deprived of normal visual input during a critical period early in development, neurons that would normally respond to that eye fail to develop properly and instead respond primarily to the other eye.

In classic experiments, Hubel and Wiesel sutured shut one eye of newborn kittens for several weeks before reopening it. After eye opening, the previously deprived eye could see normally in terms of basic light detection, but the cortical neurons that should have responded to that eye had been permanently reassigned to process input from the experienced eye. The deprived eye was functionally blind despite being structurally normal, a condition called amblyopia. Importantly, the same deprivation in adult cats had minimal effects, demonstrating that the critical period for visual development occurs early in life.

In humans, similar critical periods exist for visual development. Children born with cataracts that obscure vision must have them removed early in life to develop normal vision. If cataracts are not removed until later childhood, visual abilities remain permanently impaired even after the cataracts are removed and the eyes can transmit visual information normally. The critical period for human visual development extends from birth through approximately age seven or eight, with earlier treatment producing better outcomes.

Language development shows similar critical period effects. Children exposed to language from birth develop native fluency effortlessly. Children exposed to their first language during early childhood, even starting at age three or four, can still develop native-like abilities. However, individuals first exposed to language after early childhood, such as deaf children who did not receive sign language exposure until later, rarely achieve fully native-like language abilities even with intensive instruction.

The case of Genie, a girl discovered at age thirteen after years of severe isolation and language deprivation, tragically demonstrated critical period effects in humans. Despite intensive language instruction after her rescue, Genie never developed normal grammatical abilities, though she did acquire substantial vocabulary. Her case suggests that the critical period for acquiring grammatical aspects of language may close during childhood, making later acquisition extremely difficult.

However, critical periods are not rigidly closed windows. Some degree of plasticity persists beyond traditional critical periods, allowing learning to continue throughout life, though typically requiring more intensive training to achieve more limited results than would occur during the critical period. Additionally, recent research has identified molecular mechanisms that open and close critical periods, raising the possibility that critical periods might be artificially reopened in adulthood through pharmacological or other interventions.

The mechanisms underlying critical periods involve multiple factors. During critical periods, the brain produces high levels of growth factors and other molecules that promote plasticity. Inhibitory circuits in the cortex are still maturing, with the balance between excitation and inhibition shifting toward stronger inhibition as critical periods close. The extracellular matrix, a scaffold of proteins surrounding neurons, becomes denser and more stable, constraining structural changes. Epigenetic modifications alter gene expression in ways that reduce plasticity genes while increasing stability genes.

Manipulating these mechanisms can extend or reopen critical periods experimentally. Reducing inhibitory neurotransmission, degrading the extracellular matrix, administering certain growth factors, or altering epigenetic marks can all enhance plasticity in adult animals and enable learning that would normally be restricted to critical periods. These findings suggest that reduced adult plasticity is not simply a passive consequence of aging but actively maintained by molecular brakes that could potentially be released to facilitate learning or recovery from injury.

Neurogenesis: Birth of New Neurons in the Adult Brain

For most of the twentieth century, a central dogma of neuroscience held that neurogenesis, the birth of new neurons, occurs only during development and ceases in adulthood. This belief shaped thinking about brain injury and disease, suggesting that neurons lost to stroke, trauma, or degeneration could never be replaced. The discovery that new neurons continue to be generated in certain brain regions throughout adult life fundamentally challenged this dogma and opened new possibilities for brain repair.

The first convincing evidence for adult neurogenesis came from studies of songbirds in the 1980s. Fernando Nottebohm and colleagues discovered that male canaries generate new neurons in brain regions controlling song production, with neurogenesis rates varying seasonally in coordination with song learning. This discovery challenged the no-new-neurons dogma but was initially dismissed as a peculiarity of birds rather than a general principle applying to mammals.

However, subsequent research demonstrated adult neurogenesis in mammals including rats, mice, monkeys, and eventually humans. The primary site of adult neurogenesis is the hippocampus, specifically a region called the dentate gyrus. Here, neural stem cells continue dividing throughout life, producing new neurons that migrate short distances, extend dendrites and axons, form synaptic connections, and become integrated into existing neural circuits. Thousands of new neurons are born in the hippocampus daily, though many die before becoming fully functional neurons.

Evidence for neurogenesis in adult humans came initially from studies using radioactive tracers that label dividing cells. These studies, conducted with cancer patients who received radioactive compounds for medical purposes, demonstrated that new neurons continue to be generated in the adult human hippocampus at rates comparable to those in other mammals. More recent studies using different techniques have confirmed ongoing neurogenesis in humans, though rates appear to decline substantially with aging.

The functional significance of adult neurogenesis remains incompletely understood but appears to involve learning and memory. New neurons in the hippocampus show enhanced plasticity compared to mature neurons, with more easily modified synapses that may be particularly important for encoding new memories while minimizing interference with existing memories. Blocking neurogenesis in animals impairs certain types of learning, particularly pattern separation, the ability to distinguish similar but distinct experiences.

Numerous factors influence neurogenesis rates. Physical exercise, particularly aerobic exercise, substantially increases neurogenesis in animal studies and may do so in humans as well. Learning and environmental enrichment also promote neurogenesis. Chronic stress and elevated cortisol suppress neurogenesis, potentially contributing to stress-related cognitive impairment and depression. Antidepressant medications increase neurogenesis, and this effect may contribute to their therapeutic benefits, though this remains controversial.

The discovery of adult neurogenesis raised hopes that it might be harnessed to treat neurodegenerative diseases or repair brain injuries. If the brain retains capacity to generate new neurons, perhaps this capacity could be enhanced to replace neurons lost to Alzheimer's disease, Parkinson's disease, or stroke. However, realizing this potential has proven challenging. Neurogenesis is restricted to a few brain regions, primarily the hippocampus and olfactory bulb, and does not occur in most cortical areas where neurodegenerative damage often occurs.

Attempts to promote neurogenesis as therapy have had limited success. While various interventions can increase neurogenesis in the hippocampus, this has not translated into meaningful functional recovery in disease models. The newly born neurons must not only survive but also integrate appropriately into circuits and contribute to function, processes that appear difficult to control. Additionally, some research questions whether neurogenesis persists at meaningful levels in adult humans, with some studies suggesting it may be minimal or absent in humans beyond adolescence, though this remains debated.

Despite these limitations, the discovery of adult neurogenesis fundamentally changed understanding of brain plasticity. It demonstrated that the adult brain retains stem cells capable of generating new neurons under appropriate conditions. It showed that brain structure is even more dynamic than previously recognized, with not just synaptic connections but entire neurons being added and removed throughout life. It provided concrete evidence that the adult brain is not a static organ but a dynamic structure capable of profound change.

Plasticity in Practice: Skill Learning and Expertise

The mechanisms of neuroplasticity become concrete when examining how the brain changes during skill acquisition. Learning any complex skill, whether playing an instrument, speaking a language, or mastering a sport, involves systematic changes in brain structure and function that accumulate over months and years of practice. Understanding these changes reveals both how expertise develops and what conditions optimize learning.

The acquisition of motor skills provides clear examples of practice-induced plasticity. When someone learns to play piano, their brain undergoes multiple changes. Initially, learning requires substantial conscious attention and cognitive control, heavily engaging prefrontal and parietal regions involved in attention and executive function. Movements are slow, effortful, and error-prone. With practice, performance becomes faster, smoother, and more accurate as motor representations in primary motor cortex and premotor regions become more refined and efficient.

As skills become well-learned, control shifts from cortical regions requiring attention to subcortical structures including the basal ganglia and cerebellum that support automatic, procedural execution. The prefrontal regions active during initial learning show reduced activity as skills become automatic, freeing attention for other tasks. This is why experienced drivers can navigate familiar routes while holding conversations, whereas learners must devote full attention to driving. The transfer from cortical to subcortical control reflects a general principle: early learning is attention-demanding and cortically-dependent, while skilled performance becomes automatic and subcortically-controlled.

The "ten-thousand hour rule," popularized by Malcolm Gladwell based on research by Anders Ericsson, suggests that achieving expert-level performance in any domain requires approximately ten thousand hours of deliberate practice. While the specific number is less important than the general principle, research clearly shows that expertise requires extensive practice, typically spanning many years. Moreover, not all practice is equally effective. Deliberate practice that focuses on weaknesses, seeks constant improvement, and includes immediate feedback produces better learning than simple repetition.

Brain imaging studies of experts reveal structural and functional differences from novices. London taxi drivers, who must memorize thousands of streets and routes, show enlarged posterior hippocampi that correlate with years of experience. Professional musicians show enhanced connectivity between motor and auditory regions, reflecting tight integration of movement and sound. Bilinguals show structural differences in regions involved in executive control and language processing. Chess masters show different patterns of brain activation when viewing chess positions compared to novices, with experts showing more efficient processing in fewer brain regions.

These expert brain differences could reflect either innate advantages that predispose people toward particular domains, or plastic changes resulting from extensive practice, or some combination. Longitudinal studies that follow learners over time provide the clearest evidence that practice drives brain changes. Studies of medical students show structural changes in brain regions supporting spatial and verbal memory during intensive study periods.

Studies of people learning to juggle show expansion of regions involved in visual motion processing, with expansion reversing when practice stops. These findings demonstrate that brain structure changes in response to practice rather than simply differing between people who choose different activities.

However, individual differences in learning rates and ultimate achievement clearly exist. Some people acquire skills more quickly or reach higher levels of performance than others despite similar amounts of practice. These differences likely reflect variations in baseline brain structure and function, genetic factors affecting plasticity mechanisms, motivational factors influencing how much and how effectively people practice, and complex gene-by-environment interactions. The relative contributions of innate talent versus practice remain debated, but evidence increasingly suggests that both matter, with extensive deliberate practice being necessary but not always sufficient for elite performance.

Understanding plasticity has practical implications for optimizing learning. Distributed practice, spreading learning sessions over time with breaks between sessions, typically produces better long-term retention than massed practice, cramming all practice into concentrated periods. This spacing effect reflects advantages of allowing consolidation to occur between practice sessions and the benefits of repeated retrieval from memory.

Interleaving practice of multiple related skills, rather than blocking practice of one skill at a time, often enhances learning despite making initial practice feel more difficult. Interleaving forces learners to repeatedly retrieve and distinguish between different skills, creating stronger and more flexible learning. Blocked practice feels easier because each skill is repeated immediately while it is fresh in memory, but this reduces the challenge and limits long-term learning.

Variability in practice conditions, such as practicing a skill from different positions, in different contexts, or with different parameters, creates more robust learning that transfers better to novel situations. Constant practice under identical conditions creates learning specific to those conditions that may not generalize. This principle explains why practicing basketball free throws from exactly the same spot is less effective than practicing from various positions.

Feedback is crucial for learning but the optimal timing and nature of feedback depends on the learning stage and type of skill. Immediate feedback during early learning helps correct errors before they become habitual. Delayed feedback during later learning encourages learners to self-evaluate and self-correct, promoting deeper understanding and more independent performance. Feedback should focus on processes and strategies rather than simply outcomes, helping learners understand not just what went wrong but why and how to improve.

Sleep's importance for learning extends beyond memory consolidation discussed in the previous chapter to include motor skill learning and procedural memory. Performance on motor tasks often improves overnight without additional practice, a phenomenon called offline learning. This improvement appears to reflect sleep-dependent consolidation and refinement of motor memories, with different sleep stages contributing differently to different types of learning.

Plasticity and Recovery: Healing the Injured Brain

Perhaps the most clinically important manifestation of neuroplasticity is the brain's capacity to reorganize after injury, sometimes allowing substantial recovery of function that was initially lost. Understanding the mechanisms and limits of recovery-related plasticity has transformed rehabilitation approaches and provided hope for people affected by stroke, traumatic brain injury, and other neurological conditions.

Stroke, the sudden loss of brain function due to interrupted blood supply, affects millions of people annually and is a leading cause of adult disability. When brain tissue is deprived of blood, neurons begin dying within minutes, creating a core region of permanent damage. Surrounding this core is a penumbra of tissue that is compromised but potentially salvageable if blood flow is restored quickly. Modern stroke treatment focuses on rapidly restoring blood flow to save penumbral tissue, but once neurons in the core region die, they are permanently lost.

However, lost function is not always permanently lost. Many stroke survivors show substantial functional recovery in the weeks and months following stroke, regaining abilities that were initially severely impaired. This recovery reflects neuroplasticity, with surviving brain regions taking over functions previously performed by damaged areas. The extent and nature of recovery depends on multiple factors including the size and location of damage, the age and overall health of the patient, and the intensity and timing of rehabilitation.

Several mechanisms contribute to recovery after stroke. In the acute period immediately following stroke, recovery may partly reflect resolution of temporary dysfunction in penumbral tissue that was compromised but not destroyed. This diaschisis recovery, where temporarily suppressed regions resume function, can produce rapid improvements in the first days and weeks.

Beyond this acute recovery, more gradual improvements reflect true plasticity and reorganization. Surviving neurons in regions adjacent to the damage may sprout new connections that partially compensate for lost circuitry. The undamaged hemisphere may take over some functions previously performed by the damaged hemisphere, though this interhemispheric reorganization appears less effective than perilesional reorganization near the damage site. Behavioral compensation, where patients learn alternative strategies to accomplish tasks, also contributes to functional improvement even without true neurological recovery.

Rehabilitation approaches increasingly leverage understanding of plasticity mechanisms to promote recovery. Constraint-induced movement therapy for stroke patients with arm paralysis involves restraining the functional arm to force use of the impaired arm through intensive, repetitive practice. This approach, based on animal research showing that forced use of impaired limbs promotes cortical reorganization, can produce substantial improvements even years after stroke, challenging earlier beliefs that recovery windows closed within months.

The intensity and repetitiveness of practice appears crucial for driving plasticity. The principle of "use it or lose it" applies to damaged brains as surely as to healthy ones. Brain regions that are not used following injury fail to reorganize effectively and may lose additional function. Conversely, intensive, repetitive practice of impaired functions drives reorganization and promotes recovery. This explains why intensive rehabilitation produces better outcomes than less demanding approaches, and why recovery potential extends far beyond traditional recovery windows when appropriate interventions are provided.

Timing of rehabilitation also matters. While intensive practice appears beneficial at almost any point after stroke, earlier intervention may be more effective in some cases by shaping reorganization as it occurs. However, very early intensive practice immediately after stroke may sometimes be harmful by increasing metabolic demands in compromised tissue. Balancing early intervention to capitalize on plasticity with allowing sufficient time for acute damage processes to resolve remains an active area of investigation.

Pharmacological augmentation of plasticity represents an emerging approach to enhancing rehabilitation. Drugs that enhance neurotransmitter function, promote neurogenesis, or reduce plasticity-limiting factors can enhance recovery in animal models when combined with rehabilitation. Several compounds are being tested in humans, with some showing modest benefits.

The rationale is that drugs might open windows of enhanced plasticity during which rehabilitation would be more effective, amplifying behavioral training benefits.

Non-invasive brain stimulation techniques including transcranial magnetic stimulation and transcranial direct current stimulation can modulate cortical excitability and may enhance plasticity when combined with training. These approaches typically involve applying stimulation before or during rehabilitation sessions to prime the brain for learning. While results are mixed and optimal protocols remain under investigation, some studies suggest benefits for motor and language recovery after stroke.

Traumatic brain injury, caused by blows to the head from accidents or violence, presents different challenges than stroke. Damage is often diffuse rather than focal, affecting white matter tracts and producing widespread circuit disruptions. Recovery mechanisms are similar to those following stroke, though outcomes are highly variable depending on injury severity and distribution. Mild traumatic brain injury, commonly called concussion, typically resolves completely within weeks to months, though repeated concussions can cause cumulative damage. Severe traumatic brain injury often leaves lasting impairments, though some recovery can occur through plasticity-based reorganization.

Spinal cord injury presents particularly challenging recovery problems because damaged axons in the adult central nervous system typically fail to regenerate. While mechanisms exist for local circuit reorganization, recovery of function requiring regrowth of severed axons over long distances rarely occurs. This is why spinal cord injuries often produce permanent paralysis and sensory loss. Research into promoting axon regeneration and bridging spinal cord injuries remains active, with some promising approaches including cellular therapies, bioengineered scaffolds to guide regrowth, and pharmacological treatments to overcome growth-inhibiting factors. However, translating these approaches into effective human treatments has proven extremely difficult.

Understanding limits of plasticity is as important as recognizing its potential. Not all functions can be taken over by other brain regions. Some reorganization may be maladaptive, as when phantom limb pain results from reorganization following amputation. Recovery is typically incomplete, with residual deficits persisting even after good rehabilitation. Plasticity mechanisms that can partially compensate for small lesions may be overwhelmed by large injuries. Realistic expectations about recovery potential must be balanced with maximizing what plasticity can achieve through optimal rehabilitation.

Conclusion: The Changing Brain

The discovery that the brain remains plastic throughout life has fundamentally transformed neuroscience and its applications. It has replaced the view of the brain as a fixed structure with understanding of the brain as a dynamic organ constantly reshaped by experience. This shift has profound implications for how we approach education, rehabilitation, skill development, and mental health.

Understanding plasticity reveals that abilities are not fixed by genetics or early development but can be cultivated through appropriate practice and experience. It shows that recovery from brain injury is possible through mechanisms that recruit surviving regions to compensate for damage. It demonstrates that age-related cognitive decline is not inevitable but can be postponed through cognitively stimulating activities and healthy lifestyles. It suggests that mental health conditions involving dysfunctional neural circuits might be addressed through experiences and therapies that promote beneficial reorganization.

However, plasticity also has limits. Not all brain regions show equal plasticity.

Critical periods create windows when certain types of learning occur most readily. Recovery from severe brain damage is usually incomplete. Unwanted plasticity can occur, as when chronic pain promotes maladaptive changes in pain processing circuits. Understanding both the possibilities and limitations of plasticity allows realistic expectations and targeted interventions.

The mechanisms of plasticity, from synaptic strengthening to cortical reorganization to neurogenesis, operate continuously throughout life, driven by our experiences, behaviors, and environments. This means that we are constantly, though usually unconsciously, shaping our own brains through how we spend our time and what we practice. This places both power and responsibility in our hands: the power to cultivate desired abilities and the responsibility to provide developing brains with experiences that promote healthy development.

11

INTELLIGENCE – WHAT MAKES US SMART?

Few concepts in psychology are more controversial or more consequential than intelligence. We all have intuitions about what it means to be intelligent, and we constantly make judgments about our own intelligence and that of others. Educational systems sort students based on perceived intellectual abilities. Employers hire based partly on assessments of cognitive capacity. Social hierarchies often align with perceived intelligence. Yet despite its pervasive influence on human life, intelligence remains poorly understood and hotly debated.

What exactly is intelligence? Is it a single general ability that influences performance across all cognitive domains, or a collection of independent abilities that vary separately? How much of intelligence reflects genetic inheritance versus environmental influences? Can intelligence be meaningfully measured, and if so, what do those measurements tell us? Do intelligence tests measure something real and important, or do they reflect cultural biases that advantage some groups while disadvantaging others? Can intelligence be increased through training and education, or is it relatively fixed?

These questions are not merely academic. Beliefs about intelligence shape educational policies, influence hiring practices, affect how parents raise children, and impact how people view their own potential. The history of intelligence research includes both genuine scientific insights and terrible misuses including justifications for eugenics, racial discrimination, and educational tracking that denied opportunities to disadvantaged groups. Understanding intelligence requires grappling with this fraught history while pursuing genuine understanding of human cognitive abilities.

From a biological perspective, intelligence reflects the brain's capacity for complex information processing, problem-solving, learning, and adaptation to novel situations. This capacity depends on multiple factors including brain structure and connectivity, neural processing efficiency, neurotransmitter function, and how effectively different brain regions communicate and coordinate. Intelligence also depends on knowledge and skills acquired through experience, making it a product of both biological endowment and environmental opportunity.

This chapter explores intelligence from multiple perspectives, examining theories of what intelligence is, how it is measured, what brain features correlate with intelligence, genetic and environmental contributions, whether intelligence can be enhanced, and how understanding intelligence can inform education and policy while avoiding past mistakes. We will see that intelligence is both biologically grounded and environmentally shaped, both measurable and multifaceted, both somewhat stable and potentially malleable.

Defining and Measuring Intelligence

The scientific study of intelligence began in the late nineteenth and early twentieth centuries with attempts to develop objective measures of mental abilities. French psychologist Alfred Binet was commissioned by the French government to create a test that could identify children needing special educational assistance. Binet developed a series of tasks assessing various mental abilities including memory, attention, problem-solving, and comprehension. He introduced the concept of mental age, comparing a child's performance to the typical performance of children at different ages.

Binet's work was imported to the United States where psychologist Lewis Terman adapted and expanded it, creating the Stanford-Benet Intelligence Scale. Terman introduced the intelligence quotient or IQ, calculated as mental age divided by chronological age and multiplied by 100. A child whose mental age matched their chronological age would have an IQ of 100, considered average. A child performing like older children would have an IQ above 100, while a child performing like younger children would have an IQ below 100.

Modern IQ tests no longer use the mental age concept but instead compare performance to age peers using statistical norms. IQ scores are standardized so that the average score is 100 and most people score between 85 and 115, with fewer people scoring very high or very low. This creates a bell curve or normal distribution of scores in the population. An IQ of 130, two standard deviations above the mean, is achieved by only about two percent of people, while an IQ of 70, two standard deviations below the mean, occurs in roughly two percent of people.

What do IQ tests actually measure? Modern tests include diverse subtests assessing various cognitive abilities. The Wechsler scales, the most widely used IQ tests, include subtests measuring verbal comprehension, perceptual reasoning, working memory, and processing speed. Verbal comprehension items assess vocabulary, general knowledge, and understanding of similarities and differences. Perceptual reasoning items involve puzzles, pattern completion, and visual-spatial reasoning. Working memory items require holding and manipulating information. Processing speed items measure how quickly simple cognitive tasks can be performed.

A fundamental question is whether these diverse abilities reflect a single underlying general intelligence or are better understood as separate abilities.

Charles Spearman, an early pioneer in intelligence research, proposed that performance across different cognitive tasks shows positive correlations, meaning that people who do well on one type of task tend to do well on others. Spearman called this common factor general intelligence or g factor. He proposed that g reflects a general mental energy or capacity that influences performance across all cognitive domains, while specific abilities contribute to performance on particular tasks.

Statistical analyses consistently support the existence of g. When people complete batteries of diverse cognitive tests, their scores on different tests correlate positively. Factor analysis, a statistical technique for identifying common factors underlying correlated variables, typically extracts a general factor that accounts for substantial variance in test performance. This suggests that while people certainly have varying profiles of specific strengths and weaknesses, there is also a general cognitive ability influencing overall intellectual performance.

However, recognizing g does not mean intelligence is unitary or that specific abilities are unimportant. Raymond Cattell distinguished between fluid intelligence, the capacity to reason and solve novel problems independent of acquired knowledge, and crystallized intelligence, accumulated knowledge and skills acquired through experience and education. Fluid intelligence involves abstract reasoning, pattern recognition, and problem-solving in unfamiliar contexts. Crystallized intelligence involves vocabulary, general knowledge, and expertise in specific domains.

These two types of intelligence show different developmental trajectories. Fluid intelligence typically peaks in young adulthood and gradually declines with age, reflecting changes in processing speed and working memory. Crystallized intelligence tends to increase or remain stable into older adulthood as knowledge continues accumulating. This dissociation suggests they reflect partly different cognitive systems and brain mechanisms, though both contribute to overall intelligent behavior.

Howard Gardner proposed a theory of multiple intelligences suggesting that traditional IQ tests measure only a narrow range of abilities while ignoring others that are equally important. Gardner identified several distinct intelligences including linguistic, logical-mathematical, spatial, musical, bodily-kinesthetic, interpersonal, intrapersonal, and naturalistic intelligence. According to this theory, people have different profiles across these intelligences, with someone being highly intelligent in one domain while being average or below average in others.

While Gardner's theory has been influential in education, it has been criticized on empirical grounds. The proposed separate intelligences tend to correlate positively with each other, suggesting they are not entirely independent. Many of the proposed intelligences, like musical or bodily-kinesthetic ability, might be better characterized as talents or skills rather than forms of intelligence. The theory has been difficult to test empirically because Gardner did not develop objective measures of the proposed intelligences.

Robert Stern berg proposed a triarchic theory of intelligence distinguishing between analytical intelligence, the abilities measured by traditional IQ tests, creative intelligence, the ability to deal with novel situations and generate innovative ideas, and practical intelligence, the ability to adapt to real-world environments and solve everyday problems. Stern berg argued that traditional tests overemphasize analytical intelligence while neglecting creative and practical abilities that are crucial for real-world success.

Emotional intelligence, proposed by Peter Salovey and John Mayer and popularized by Daniel Goleman, refers to the ability to perceive, understand, manage, and use emotions effectively. This includes recognizing emotions in oneself and others, understanding how emotions influence thinking and behavior, regulating emotional responses, and using emotional information to guide decisions and social interactions. Emotional intelligence appears to predict important life outcomes including relationship quality, mental health, and job performance in positions requiring interpersonal skills.

The proliferation of intelligence theories reflects both genuine insights about the multifaceted nature of human abilities and disagreements about what should count as intelligence versus related but distinct constructs. Most researchers agree that human cognitive abilities are complex and multifaceted, that people show varying profiles of strengths and weaknesses, and that traditional IQ tests measure important but not all aspects of cognition. The debate centers on whether it is more useful to emphasize general intelligence that influences overall performance versus specific abilities that operate independently, and whether constructs like creativity, practical skills, and emotional abilities should be considered types of intelligence or separate constructs.

The Biology of Intelligence

Understanding intelligence requires understanding its biological substrates in the brain. What brain characteristics distinguish more intelligent from less intelligent individuals? Researchers have investigated brain size, structure, connectivity, neural efficiency, and neurochemistry in relation to intelligence, revealing that multiple biological factors contribute to cognitive ability.

The most obvious brain characteristic that might relate to intelligence is size. Across species, brain size relative to body size correlates with behavioral complexity, with humans having exceptionally large brains relative to body size compared to other animals. Within humans, does brain size correlate with intelligence? Early attempts to answer this question through crude measurements of skull size produced inconsistent and often biased results. Modern brain imaging provides more accurate measurements and shows modest positive correlations between brain volume and IQ, with correlation coefficients typically around 0.3 to 0.4.

This correlation means that about 10 to 15 percent of variance in intelligence is associated with brain size. Brain size clearly matters to some extent, but it is far from the whole story, with most variation in intelligence unrelated to size. Moreover, the correlation does not prove that larger brains cause higher intelligence. Both could reflect common causes such as better nutrition and health during development.

More important than overall brain size appears to be the size and organization of specific brain regions, particularly those involved in higher cognitive functions. The prefrontal cortex, which supports executive functions including working memory, planning, and cognitive control, shows particularly strong associations with intelligence. Intelligence correlates with larger prefrontal cortex volume and with greater thickness of prefrontal cortex gray matter. Parietal regions involved in spatial reasoning and attention also show associations with intelligence.

Beyond regional volumes, the pattern of connectivity between brain regions appears crucial for intelligence. Higher intelligence is associated with more efficient long-range connections linking distributed brain regions into coordinated networks. This suggests that intelligence depends not just on local processing power within individual regions but on how effectively different regions communicate and integrate information.

The parieto-frontal integration theory, proposed by neuroscientist Rex Jung and colleagues, suggests that intelligence emerges from effective integration of information processing across a network of frontal and parietal regions. This network, sometimes called the multiple-demand network because it activates across diverse cognitively demanding tasks, appears to be particularly important for fluid intelligence and novel problem-solving.

Neural efficiency also correlates with intelligence. More intelligent individuals often show less brain activation when performing cognitive tasks, particularly after practice, suggesting their brains work more efficiently and require less effort to achieve the same performance. This has been observed across various tasks and brain regions, supporting the neural efficiency hypothesis that intelligent brains process information more efficiently rather than simply working harder.

Processing speed, how quickly the brain can process information, shows moderate correlations with intelligence. More intelligent individuals tend to perform simple cognitive tasks faster, show faster reaction times, and have faster nerve conduction velocities. These associations suggest that basic neural processing efficiency contributes to higher-level cognitive abilities, though processing speed alone cannot account for intelligence.

Working memory capacity, the ability to hold and manipulate information temporarily, shows strong correlations with intelligence, particularly fluid intelligence. Individual differences in working memory capacity account for substantial variance in IQ scores, suggesting that the ability to maintain and operate on information is central to intelligent thought. Brain regions supporting working memory, particularly prefrontal and parietal cortex, show activity patterns that differ between high and low working memory capacity individuals.

White matter integrity, reflecting the quality of connections between brain regions, correlates with intelligence. Diffusion tensor imaging, a technique that measures water diffusion along axon bundles, reveals that more intelligent individuals tend to have higher fractional anisotropy, a measure of white matter organization and myelination quality. Better organized, well-myelinated connections presumably allow faster and more reliable information transfer between regions.

Neurotransmitter function also influences intelligence, though this is less well understood than structural factors. Dopamine appears particularly important, with dopamine receptor density and dopamine signaling efficiency showing associations with working memory and fluid intelligence. The inverted-U relationship between dopamine levels and cognitive performance suggests that optimal dopamine function, neither too little nor too much, supports best performance.

The concept of neural plasticity discussed in the previous chapter also relates to intelligence. More intelligent individuals may show greater capacity for learning and neural adaptation, allowing them to acquire new knowledge and skills more readily. However, the relationship between plasticity and intelligence is complex, with some evidence that more intelligent individuals show more selective plasticity, strengthening important information while efficiently pruning irrelevant information.

Genetic influences on brain structure and function contribute to the heritability of intelligence, discussed in the next section. Genes affecting neuron growth, synapse formation, neurotransmitter function, myelination, and neural migration all potentially influence intelligence by shaping brain development and function. However, these genetic influences interact with environmental factors including nutrition, education, and cognitive stimulation, making intelligence a product of both biological endowment and environmental opportunity.

Nature and Nurture: Genes, Environment, and Intelligence

Few questions in psychology have generated more controversy than the relative contributions of genetic inheritance and environmental factors to intelligence. The debate often becomes politicized, with some emphasizing genetic influences to argue that intervention cannot substantially improve outcomes, while others emphasize environmental factors to argue that all differences reflect opportunity rather than ability. The scientific evidence supports a nuanced view: both genes and environment substantially influence intelligence, they interact in complex ways, and understanding their contributions has important implications without supporting either genetic determinism or environmental determinism.

Behavioral genetics research using twin studies, adoption studies, and more recently molecular genetics has consistently found that intelligence is heritable, meaning that genetic differences among individuals account for substantial variation in IQ scores. Heritability estimates for intelligence in adults typically range from 50 to 80 percent, meaning that roughly half to three-quarters of variation in intelligence within populations reflects genetic differences while the remainder reflects environmental differences and measurement error.

However, heritability is commonly misunderstood. Heritability describes what proportion of variation in a population is associated with genetic variation, not what proportion of an individual's intelligence is determined by genes versus environment. Every individual's intelligence reflects both genetic and environmental influences working together. Heritability also does not indicate whether differences between groups reflect genetic or environmental factors. High heritability within groups tells us nothing about causes of differences between groups.

Moreover, heritability is not fixed but can vary across populations and environments. In environments where everyone has similar educational opportunities and resources, environmental variation is minimized and heritability increases because most remaining variation reflects genetic differences. In environments with large disparities in opportunity, environmental variation is substantial and heritability decreases. Studies consistently find that heritability of intelligence is higher in affluent families and communities compared to disadvantaged ones, suggesting that when basic needs are met and opportunities are broadly available, genetic differences become more influential, while in harsh or deprived environments, environmental factors dominate.

This gene-environment interaction has important implications. It suggests that reducing environmental disparities might allow genetic potential to be more fully realized across all groups. It challenges the notion that high heritability means intervention is futile, showing instead that improving environments can reduce the impact of genetic differences.

Molecular genetics has begun identifying specific genes influencing intelligence, though progress has been slower than for other heritable traits. Genome-wide association studies comparing genetic variants across thousands of individuals have identified hundreds of genetic variants associated with intelligence, but each variant has tiny effects, typically accounting for less than one percent of variance. The largest studies have identified variants collectively explaining about 10 to 15 percent of variance in intelligence, still far less than the 50 to 80 percent heritability estimated from twin studies.

This missing heritability likely reflects multiple factors. Many more genetic variants with even smaller individual effects likely contribute, requiring even larger studies to detect. Gene-by-gene interactions, where effects of one gene depend on variants in other genes, add complexity. Gene-by-environment interactions, where genetic effects depend on environmental contexts, create additional complexity that standard analyses may miss. Rare variants not captured by common variant studies may contribute. Some heritability may reflect indirect genetic effects, where parental genes influence parenting behaviors and home environments that in turn affect children's intelligence.

Environmental influences on intelligence are substantial and diverse. Prenatal environment influences brain development, with factors including maternal nutrition, stress, toxin exposure, and infections affecting fetal brain growth. Children born to malnourished mothers or exposed prenatally to alcohol, lead, or other toxins show reduced cognitive abilities. Premature birth and low birth weight predict lower intelligence, reflecting disrupted brain development.

Early childhood environment powerfully shapes cognitive development. Severe deprivation, as experienced by children in orphanages with minimal stimulation and interaction, produces significant cognitive delays. Children adopted from severely deprived institutions into nurturing families often show dramatic improvements, though some deficits may persist, particularly when deprivation extended beyond early childhood. This demonstrates both the importance of early environmental input and the brain's capacity for recovery through plasticity.

Quality of parenting and home environment influences intelligence. Children experiencing responsive, sensitive parenting, rich language environments with extensive conversation, access to books and educational materials, and cognitive stimulation through play and activities show higher cognitive development. Parental education and socioeconomic status correlate with child intelligence, partly through genetic transmission but also through environmental mechanisms including differences in home cognitive environments.

Education is perhaps the most important environmental influence on intelligence. Each additional year of schooling produces measurable increases in IQ, with estimates typically around 1 to 5 IQ points per year of education. This effect persists after controlling for initial ability and family background, suggesting education causes increases rather than simply reflecting pre-existing differences. The effects of education on intelligence demonstrate that intelligence is not fixed but can be systematically increased through appropriate interventions.

Nutrition affects cognitive development and function. Severe malnutrition during critical developmental periods causes permanent cognitive impairment. More moderate nutritional inadequacies, including deficiencies in iron, iodine, and other micronutrients, impair cognitive function. Supplementation studies show that providing adequate nutrition to deficient populations improves cognitive performance, particularly when intervention occurs early in development.

Environmental toxins impair cognitive development, with lead exposure being the most thoroughly documented. Even low levels of lead exposure reduce IQ, with no apparent safe threshold. The removal of lead from gasoline and paint has been associated with population-level IQ increases, demonstrating how environmental changes can systematically affect intelligence. Other environmental toxins including mercury, pesticides, and air pollution also appear to impair cognitive function.

The Flynn effect, named after researcher James Flynn who documented it, refers to substantial increases in IQ scores across generations in many countries. Average IQ scores have increased by about 3 points per decade throughout the twentieth century in developed nations, too rapid to reflect genetic changes and therefore clearly reflecting environmental improvements. Multiple factors likely contribute including better nutrition, reduced infectious disease, more education, increased cognitive complexity of modern environments, and greater test-taking familiarity.

However, the Flynn effect appears to have slowed or reversed in some countries in recent decades, with IQ scores plateauing or even declining. This may reflect reaching ceilings on environmental improvements that previously drove gains, or new environmental challenges including digital media effects on attention and reading. Understanding factors that promote or impair cognitive development remains an active area of research.

Can Intelligence Be Increased?

Given that intelligence reflects both genetic and environmental influences, and that environmental factors clearly affect cognitive development, can intelligence be systematically increased through training or intervention? This question has enormous practical importance for education and policy, motivating extensive research on cognitive training programs.

Early intervention programs targeting disadvantaged children have shown the most consistent success in improving cognitive outcomes. The Perry Preschool Project, a landmark study begun in the 1960s, provided high-quality preschool education to disadvantaged children and followed them for decades. The program produced initial IQ gains that faded over time, but participants showed lasting benefits including higher graduation rates, employment, and earnings, and lower crime rates. This suggests that early intervention can produce lasting benefits even when initial IQ gains are not sustained.

The Abecedarian Project provided intensive educational intervention starting in infancy and continuing through age five to children from disadvantaged backgrounds. This intervention produced sustained IQ gains that persisted into adulthood, with treated individuals scoring about 5 points higher than controls even in their thirties. Participants also showed better educational and employment outcomes. These results demonstrate that intensive, high-quality early intervention beginning in infancy can produce lasting cognitive benefits.

Head Start, a large-scale preschool program serving disadvantaged children across the United States, produces modest short-term cognitive gains that typically fade when children enter elementary school. However, participants show lasting benefits in education, health, and social outcomes, suggesting that early intervention benefits extend beyond IQ scores to important life outcomes.

These findings suggest that early childhood is a particularly sensitive period when environmental enrichment can have lasting effects. The mechanisms likely include both direct effects on brain development during periods of high plasticity, and cascading effects where early learning creates foundations for subsequent learning. Children who enter school better prepared are more likely to succeed, creating positive trajectories, while those entering unprepared may fall behind and struggle to catch up.

Working memory training has been promoted as a method to improve general cognitive abilities. Programs typically involve practicing computerized working memory tasks that adapt to individual performance, gradually increasing difficulty. Some studies have found that such training improves working memory capacity and shows transfer to untrained tasks including measures of fluid intelligence. However, many studies have found limited transfer beyond practiced tasks, and meta-analyses suggest that far transfer to general intelligence is modest at best.

The debate over working memory training effectiveness highlights challenges in cognitive training research. Many studies lack adequate control groups, fail to assess lasting effects beyond immediate post-training, and do not examine whether improvements transfer to real-world outcomes. Expectations and motivation may contribute to observed benefits. The most optimistic interpretation is that intensive working memory training can produce small improvements in fluid intelligence, while more skeptical interpretations suggest observed benefits reflect practice effects specific to trained tasks.

Broader cognitive training programs aiming to improve multiple abilities have produced mixed results. Brain training programs marketed to consumers often make strong claims about improving intelligence and preventing cognitive decline, but rigorous evaluations typically find limited evidence for broad benefits. Improvements on trained tasks are common, but transfer to untrained tasks and real-world abilities is often minimal.

However, specific cognitive interventions targeting particular skills can be effective. Reading interventions improve reading abilities. Math interventions improve math skills. These domain-specific improvements may not transfer to general intelligence but are valuable in their own right. The question is whether training can produce broad improvements in general cognitive ability or whether benefits are typically limited to trained domains.

Physical exercise, particularly aerobic exercise, appears to benefit cognitive function through effects on brain health. Exercise increases blood flow to the brain, promotes neurogenesis in the hippocampus, increases production of brain-derived neurotrophic factor that supports neural health, and improves mood and sleep. Studies in children and adults show that aerobic fitness correlates with better cognitive performance, and exercise interventions improve cognitive function, particularly executive functions. These benefits appear more robust than those from cognitive training alone.

Multilingualism provides an example of cognitive benefits from sustained engagement in a complex skill. Bilingual individuals show advantages in executive functions including attention control, task switching, and interference management, likely reflecting constant practice in managing multiple languages. These benefits appear most pronounced when both languages are actively used throughout life. However, bilingualism does not appear to increase IQ per SE but rather enhances specific executive control abilities.

Musical training is associated with enhanced cognitive abilities including better working memory, attention, and processing speed. Children randomly assigned to music lessons show cognitive improvements compared to controls, suggesting causal effects. However, the size of benefits is debated, and effects on general intelligence are modest. Music training may provide cognitive benefits through sustained engagement of attention, working memory, and fine motor control, similar to benefits from other complex skill learning.

The most consistent message from intervention research is that education matters. Each year of schooling produces measurable IQ increases, accumulated educational experiences shape cognitive development, and educational improvements at the societal level contribute to population intelligence increases. Rather than searching for magic bullets that rapidly boost intelligence, the evidence suggests that sustained engagement in cognitively demanding activities including formal education produces the most reliable cognitive benefits.

The concept of growth mindset, developed by psychologist Carol Dweck, proposes that believing intelligence can be developed through effort and learning, rather than being fixed, promotes better learning outcomes. Students with growth mindsets show greater motivation to learn, better resilience when facing challenges, and ultimately better achievement. Interventions promoting growth mindsets have shown promise in improving academic outcomes, particularly for disadvantaged students.

However, mindset interventions have shown variable effectiveness, with some studies finding substantial benefits and others finding minimal effects. The conditions under which mindset interventions work best remain under investigation. Nevertheless, the basic principle that viewing intelligence as develop able rather than fixed promotes learning effort seems well-supported, even if optimal methods for instilling this view are still being refined.

Intelligence, Education, and Society

Understanding intelligence has important implications for education and social policy. How should schools respond to individual differences in cognitive abilities? Should education systems track students into different programs based on perceived ability, or should all students receive the same instruction? How much emphasis should be placed on standardized testing? What interventions can reduce achievement gaps between advantaged and disadvantaged groups?

Ability grouping and tracking, where students are sorted into different educational tracks based on measured abilities, remains controversial. Proponents argue that grouping allows instruction to be tailored to student needs, preventing advanced students from being held back and struggling students from being overwhelmed. Critics argue that tracking creates self-fulfilling prophecies, denies opportunities to students placed in lower tracks, and exacerbates inequalities by providing better instruction to already-advantaged students.

Research evidence suggests tracking produces mixed effects. High-achieving students sometimes benefit from accelerated programs, while low-achieving students rarely benefit and often suffer from lower-quality instruction and reduced expectations. Within-class ability grouping, where students remain in heterogeneous classes but work in smaller groups matched by ability for specific subjects, may provide some benefits of tailored instruction without the negative effects of rigid tracking.

Standardized testing plays a major role in education but is increasingly controversial. IQ tests and achievement tests provide objective measures of cognitive abilities and knowledge, helping identify students needing special services and tracking educational progress. However, over reliance on testing can narrow curriculum to tested content, create stress, disadvantage students from backgrounds underrepresented in test development samples, and reduce education to test preparation.

The relationship between IQ and academic achievement is substantial but imperfect. IQ predicts academic performance with correlations typically around 0.5, meaning IQ accounts for about 25 percent of variance in achievement. This is a large effect, but it leaves 75 percent of variance in achievement unexplained by IQ. Motivation, study habits, teaching quality, family support, and other factors also substantially influence achievement.

Growth mindset interventions and other approaches emphasizing effort, strategies, and persistence aim to help students maximize their potential regardless of initial ability. Teaching students effective learning strategies, providing support when they struggle, maintaining high expectations for all students, and creating environments where mistakes are learning opportunities rather than failures can improve outcomes, particularly for students who might otherwise disengage.

Reducing achievement gaps between socioeconomic and racial/ethnic groups remains a major educational challenge. These gaps appear early, before formal schooling begins, and often widen during school years. The gaps partly reflect environmental differences including nutrition, home cognitive environments, stress, neighborhood resources, and school quality. Genetic differences likely also contribute to individual variation within all groups, but evidence overwhelmingly indicates that average group differences in measured intelligence reflect environmental and cultural factors rather than genetic differences in capability.

Interventions addressing achievement gaps must target multiple levels. Early childhood interventions providing high-quality preschool to disadvantaged children help close gaps before they widen. School improvements including better teaching, higher expectations, adequate resources, and supportive climates benefit all students but particularly help disadvantaged students. Addressing broader social inequities including poverty, housing insecurity, neighborhood violence, and healthcare access is also necessary, as educational interventions alone cannot fully overcome pervasive environmental disadvantages.

The concept of stereotype threat, where awareness of negative stereotypes about one's group impairs performance, may contribute to achievement gaps. When testing situations highlight group membership and associated stereotypes, members of negatively stereotyped groups may under perform due to anxiety and divided attention. Interventions reducing stereotype threat through affirmations of values or re framing tasks can improve performance for stigmatized groups.

Beyond formal education, intelligence influences numerous life outcomes. Higher intelligence predicts better job performance, particularly for complex jobs requiring learning and problem-solving. It predicts higher income, though the relationship is moderate and many factors beyond intelligence influence earnings. It predicts better health and longer life, likely through multiple pathways including better health decisions, higher incomes allowing better healthcare, and possible shared factors influencing both intelligence and health.

However, intelligence is not destiny. Many highly intelligent people struggle due to mental illness, lack of opportunity, or other challenges. Many people of average intelligence achieve great success through motivation, persistence, social skills, and good fortune. Reducing human worth or potential to IQ scores is scientifically unjustified and morally wrong. Intelligence is one factor among many that influence life outcomes, and societies should strive to provide opportunities for all people to develop and use their abilities.

Conclusion: Intelligence in Perspective

Intelligence is real, measurable, biologically grounded, substantially heritable, and predictive of important life outcomes. These statements are scientifically well-supported based on extensive research. At the same time, intelligence is multifaceted, shaped by environment, modifiable through intervention, imperfectly measured by current tests, and only one factor among many that determine who we are and what we can achieve.

Understanding intelligence requires avoiding both extremes of genetic determinism and blank slate environmentalism. Genes matter, creating individual differences in cognitive abilities that contribute to different educational and career paths. Environment matters, shaping cognitive development in profound ways and creating opportunities for abilities to be developed or constrained. Genes and environment interact, with genetic potentials requiring supportive environments to be realized and environmental interventions being more or less effective depending on individual characteristics.

The biology of intelligence reveals that cognitive abilities depend on brain structure, connectivity, efficiency, and neurochemistry, all of which reflect complex interactions between genetic programs and environmental influences during development and throughout life. Understanding these biological foundations helps identify leverage points for intervention while respecting that biology creates both possibilities and constraints.

The practical implications of intelligence research must be pursued carefully, learning from past mistakes where intelligence testing was misused to justify discrimination, deny opportunities, and inflict suffering. Intelligence tests can be useful tools for identifying students needing support, evaluating interventions, and understanding human cognition, but they must not be used to limit human potential or excuse failure to provide opportunities.

Moving forward, the challenge is using understanding of intelligence to promote human flourishing while resisting simplistic interpretations and harmful applications. This requires recognizing both the reality of cognitive differences and the fundamental worth and potential of all people regardless of measured intelligence. It requires investing in interventions that help all people develop their abilities while acknowledging that optimal policies may need to accommodate individual differences. It requires humility about what we know and continuing research to better understand the biological and environmental factors that create human cognitive diversity.

12

SLEEP, DREAMS, AND COGNITIVE HEALTH

Every night, you surrender consciousness for approximately eight hours, entering a state of profound vulnerability during which you are largely unaware of your surroundings and unable to respond to threats. This might seem like a dangerous and wasteful allocation of time, consuming roughly one-third of your entire life in apparent inactivity. Yet sleep is not merely the absence of wakefulness but an active, highly organized biological process essential for survival. Animals deprived of sleep eventually die, often before they would die of starvation. Humans subjected to prolonged sleep deprivation experience devastating cognitive impairment, emotional instability, and eventually hallucinations and delirium.

The universality of sleep across the animal kingdom testifies to its fundamental importance. From fruit flies to elephants, virtually all animals sleep in some form, though the amount and pattern varies enormously. Dolphins and some bird species have evolved the remarkable ability to sleep with one brain hemisphere at a time, allowing them to maintain vigilance and continue swimming or flying while sleeping. The fact that evolution has preserved sleep despite its apparent costs suggests it serves vital functions that cannot be accomplished during wakefulness.

For most of human history, the functions of sleep remained mysterious. Ancient peoples viewed sleep as a kind of temporary death, a journey to other realms, or communion with gods and spirits. Even as scientific understanding advanced, sleep was long viewed as simply the brain shutting down for rest, a passive state of minimal activity. This view has been completely overturned. Modern neuroscience reveals that sleep involves complex, coordinated brain activity serving crucial functions including memory consolidation, metabolic restoration, emotional regulation, and clearance of toxic waste products from the brain.

The advent of electroencephalography (EEG), which records electrical activity from the scalp, revolutionized sleep research by allowing scientists to observe what the brain does during sleep. In 1953, Eugene Aserinsky and Nathaniel Kleitman discovered rapid eye movement (REM) sleep, a stage characterized by rapid eye movements, vivid dreams, and brain activity resembling wakefulness. This discovery shattered the notion of sleep as a uniform state of inactivity, revealing instead that sleep involves distinct stages with different characteristics and functions.

This chapter explores the biology of sleep and dreams, examining the brain systems that regulate sleep and wakefulness, the stages of sleep and their functions, how sleep supports learning and memory, the consequences of sleep deprivation, what dreams are and what purposes they might serve, sleep disorders and their treatment, and how modern lifestyle threatens healthy sleep despite its crucial importance for cognitive and physical health.

The Architecture of Sleep: Stages and Cycles

A typical night of sleep is not a single uniform state but a carefully orchestrated progression through different stages that cycle repeatedly throughout the night. Understanding this sleep architecture reveals the complexity of sleep and helps explain its multiple functions.

Sleep researchers distinguish between non-rapid eye movement (non-REM) sleep and REM sleep based on distinctive patterns of brain activity, eye movements, and muscle tone. Non-REM sleep is further divided into three stages representing increasing depth of sleep.

Stage 1 non-REM sleep is the transition between wakefulness and sleep, lasting only a few minutes. Brain waves slow from the rapid beta waves characteristic of alert wakefulness to slower alpha and theta waves. Muscle tone decreases but remains present. People awakened from stage 1 sleep may not even realize they were asleep, reporting that they were "just resting their eyes." This stage represents the lightest sleep, easily disrupted by noise or other stimuli.

Stage 2 non-REM sleep represents the onset of true sleep and comprises roughly 45 to 55 percent of total sleep time in adults. Brain waves continue slowing with occasional bursts of rapid waves called sleep spindles and large slow waves called K-complexes. These distinctive patterns appear to play roles in memory consolidation and protecting sleep from disruption. Heart rate and breathing slow, body temperature drops, and awareness of external environment diminishes. People awakened from stage 2 clearly recognize they were asleep.

Stage 3 non-REM sleep, also called slow-wave sleep or deep sleep, is characterized by large, slow delta waves dominating the EEG. This is the deepest stage of sleep, most difficult to wake from, and most important for physical restoration. Growth hormone secretion peaks during slow-wave sleep, tissue repair occurs, immune function is enhanced, and metabolic restoration takes place. Slow-wave sleep is particularly prevalent during the first half of the night and decreases with aging. Children spend much more time in slow-wave sleep than adults, consistent with their greater growth and development needs.

After progressing through the stages of non-REM sleep, the sleeper enters REM sleep, a stage radically different from non-REM sleep despite occurring during sleep. Brain activity during REM sleep resembles waking activity, with rapid, low-amplitude waves similar to those during alert wakefulness. The eyes move rapidly beneath closed lids, giving this stage its name. Most vivid, narrative dreams occur during REM sleep, though some mental activity occurs during non-REM stages as well.

Paradoxically, while the brain is highly active during REM sleep, the body is paralyzed. Muscle tone drops to its lowest levels during sleep, with voluntary muscles essentially limp except for the diaphragm controlling breathing and the eye muscles. This REM atonia appears to prevent people from physically acting out their dreams. When this paralysis mechanism fails, people can develop REM sleep behavior disorder, where they physically enact dreams, sometimes injuring themselves or bed partners.

Other physiological changes during REM sleep include irregular breathing and heart rate, increased blood pressure variability, and in males, penile erections. Body temperature regulation becomes impaired during REM sleep, making the body more vulnerable to ambient temperature extremes. These characteristics make REM sleep a unique state sometimes described as paradoxical sleep because the brain is active like waking while the body is more profoundly inactive than during deep non-REM sleep.

These sleep stages do not occur randomly but cycle in predictable patterns throughout the night. A complete sleep cycle from stage 1 non-REM through REM sleep takes approximately 90 to 110 minutes, with most people experiencing four to six complete cycles per night. However, the composition of cycles changes across the night. Early cycles contain more slow-wave sleep and shorter REM periods, while later cycles contain little or no slow-wave sleep and progressively longer REM periods. The final REM period before waking may last 30 to 60 minutes, which is why people often wake from dreams.

This changing composition of sleep cycles across the night suggests different sleep stages serve different functions, with slow-wave sleep addressing different needs than REM sleep. The front-loading of slow-wave sleep may reflect the urgency of physical restoration and metabolic processes occurring during deep sleep, while the increasing REM sleep later in the night may reflect cognitive processing functions that can occur later in the sleep period.

Brief awakenings occur throughout the night, particularly at transitions between sleep cycles, though people typically do not remember these awakenings unless they last more than a few minutes. This is why people often report waking only once or twice per night when sleep recordings show dozens of brief arousals. The ability to quickly return to sleep after such arousals is part of healthy sleep, while difficulty returning to sleep characterizes insomnia.

The Neurobiology of Sleep and Wakefulness

Sleep and wakefulness are regulated by complex brain systems involving multiple neurotransmitters, brain regions, and regulatory mechanisms. Understanding these systems reveals how sleep is controlled and what can go wrong to produce sleep disorders.

The basic pattern of sleep and wakefulness is controlled by two interacting processes called the homeostatic sleep drive and the circadian rhythm. The homeostatic sleep drive reflects the accumulation of sleep need during wakefulness. The longer you stay awake, the stronger the drive to sleep becomes, building up sleep pressure that is relieved by sleeping. This process is regulated by adenosine, a neurotransmitter that accumulates in the brain during wakefulness and promotes sleepiness by inhibiting arousal-promoting neurons. Caffeine works primarily by blocking adenosine receptors, preventing adenosine from promoting sleepiness.

The circadian rhythm is a roughly 24-hour internal biological clock that regulates sleepiness and alertness across the day-night cycle independent of how long someone has been awake. This clock is controlled by the suprachiasmatic nucleus of the hypothalamus, a small cluster of neurons that generates rhythmic patterns of activity even in isolation from external cues. The circadian clock is synchronized to the external day-night cycle primarily through light exposure detected by specialized retinal cells that project directly to the suprachiasmatic nucleus.

The interaction between homeostatic sleep drive and circadian rhythm creates the typical pattern of one consolidated sleep period at night for most humans. Sleep pressure builds across the day, and the circadian system promotes wakefulness during daylight hours but switches to promoting sleep in the evening. At night, both systems align to maintain sleep, while in the morning, the circadian system begins promoting wakefulness even though sleep pressure has been relieved.

This two-process model explains several sleep phenomena. The afternoon dip in alertness that many people experience reflects a circadian trough in arousal. The second wind some people experience late at night when they stay up past their normal bedtime reflects a circadian increase in arousal despite high sleep pressure. Jet lag occurs when the circadian system remains synchronized to the previous time zone while the new environment demands sleep and wakefulness at inappropriate circadian phases. Shift work creates similar circadian misalignment when people try to sleep during their biological day.

Multiple brain systems promote wakefulness through release of arousal-promoting neurotransmitters. The reticular activating system in the brain stem releases acetylcholine and glutamate throughout the brain. The locus coeruleus releases norepinephrine. The raphe nuclei release serotonin. The tuberomammillary nucleus releases histamine, which is why antihistamines cause drowsiness by blocking histamine receptors. The basal fore brain releases acetylcholine. These systems collectively maintain the activated brain state characteristic of wakefulness.

A particularly important arousal system involves neurons in the lateral hypothalamus that produce orexin, also called hypocretin. These neurons project widely throughout the brain and help stabilize the wake state, preventing unwanted transitions to sleep. Loss of orexin neurons causes narcolepsy, a disorder characterized by excessive daytime sleepiness and sudden attacks of muscle weakness or sleep, demonstrating the crucial role of orexin in maintaining stable wakefulness.

Sleep-promoting systems counteract these arousal systems. The ventrolateral preoptic nucleus in the hypothalamus contains neurons that become active during sleep and inhibit the arousal-promoting systems. These sleep-promoting neurons appear to be inhibited by arousal systems during wakefulness, creating a mutually inhibitory flip-flop switch where wake-promoting and sleep-promoting systems inhibit each other. This arrangement creates bi stability, making the brain tend toward either wakefulness or sleep rather than intermediate states, though it can also create instability if one side is weakened, as occurs in narcolepsy.

The transition between non-REM and REM sleep involves additional brain systems. REM sleep is actively generated by neurons in the pons, a brainstem region, that trigger the characteristic features of REM including rapid eye movements, muscle paralysis, and activated EEG. The neurotransmitter acetylcholine is particularly important for REM sleep, with cholinergic neurons in the brainstem and basal forebrain promoting REM. This explains why cholinesterase inhibitors, drugs that increase acetylcholine used to treat Alzheimer's disease, can increase REM sleep and vivid dreams.

Different neurotransmitter systems show different patterns of activity across sleep stages. Norepinephrine, serotonin, and histamine levels are high during waking, decrease during non-REM sleep, and are virtually absent during REM sleep. This withdrawal of monoamine neurotransmitters during REM may contribute to the distinctive qualities of REM dreams including their often bizarre, emotionally intense, and illogical character. Acetylcholine shows a different pattern, high during both waking and REM sleep but low during non-REM sleep.

Understanding the neurobiology of sleep has led to pharmacological approaches to sleep problems. Hypnotic medications that promote sleep typically work by enhancing GABA inhibitory neurotransmission, reducing overall brain excitability. Stimulants that promote wakefulness including caffeine, amphetamines, and modafinil work through various mechanisms including blocking adenosine, increasing dopamine and norepinephrine, and enhancing orexin function. However, medications are generally less effective than behavioral approaches for most sleep problems and carry risks of side effects and dependence.

Functions of Sleep: Why We Need to Sleep

Despite enormous research efforts, the fundamental question of why we sleep remains incompletely answered. Sleep clearly serves multiple important functions, but debate continues about which functions are most fundamental and whether sleep evolved primarily to serve one function with others being secondary benefits.

Memory consolidation is one of the best-established functions of sleep. As discussed in Chapter 9, sleep, particularly slow-wave sleep and REM sleep, plays crucial roles in stabilizing and integrating newly learned information. During sleep, the brain appears to replay recent experiences, strengthening neural connections that encode memories while pruning unnecessary connections. Sleep deprivation impairs memory formation, and many studies show that sleeping after learning improves later retention compared to remaining awake for the same period.

Different types of memory appear to benefit from different sleep stages. Declarative memories for facts and events benefit particularly from slow-wave sleep, while procedural memories for skills may benefit more from REM sleep. Emotional memories appear to be processed during REM sleep, with some research suggesting that REM sleep helps integrate emotional experiences while reducing their emotional intensity, potentially contributing to emotional regulation.

The synaptic homeostasis hypothesis, proposed by Giulio Tononi and Chiara Cirelli, suggests that one fundamental function of sleep is to downscale synaptic strengths that increase during waking as a result of learning and neural activity. During wakefulness, learning and experience strengthen synapses throughout the brain, increasing energy demands and potentially saturating capacity for further learning. During sleep, particularly slow-wave sleep, overall synaptic strengths are reduced back to sustainable levels while preserving relative changes that encode memories. This selective downscaling allows the brain to remain responsive and capable of further learning.

Metabolic restoration is another important sleep function. The brain is metabolically expensive, consuming roughly 20 percent of the body's energy despite comprising only 2 percent of body weight. During waking activity, the brain accumulates metabolic waste products and depletes energy stores. Sleep provides an opportunity for restoration, with glycogen stores being replenished and metabolic byproducts cleared.

A recently discovered function of sleep involves clearance of metabolic waste through the glymphatic system. During sleep, spaces between brain cells expand, allowing cerebrospinal fluid to flow through brain tissue more efficiently, washing away accumulated waste products including proteins like amyloid-beta that are implicated in Alzheimer's disease. This clearance function is dramatically enhanced during sleep compared to wakefulness, providing a compelling reason why adequate sleep is essential for long-term brain health.

Immune function is supported and enhanced by sleep. During sleep, the immune system ramps up production of cytokines and other immune mediators that fight infections. Sleep deprivation impairs immune responses, increasing susceptibility to infections. This is why people often sleep more when sick; the body prioritizes sleep to support immune function. The reciprocal relationship between sleep and immune function explains why infections cause fatigue and sleepiness, and why adequate sleep helps fight off illness.

Emotional regulation is profoundly affected by sleep. Sleep deprivation increases emotional reactivity, making people more irritable, anxious, and prone to negative emotions. Brain imaging studies show that sleep deprivation increases amygdala reactivity to negative stimuli while reducing prefrontal control of emotional responses. This combination creates heightened emotional responses that are poorly regulated, explaining the emotional volatility of sleep-deprived individuals.

REM sleep appears particularly important for emotional processing. The REM sleep dreaming state allows processing of emotional experiences in a context where the stress hormone norepinephrine is suppressed, potentially allowing integration of emotional memories without the physiological stress response they originally triggered. This might explain why sleep helps people cope with stressful experiences and why sleep disruption contributes to emotional disorders including depression, anxiety, and PTSD.

Cardiovascular health is supported by sleep through multiple mechanisms. Blood pressure drops during sleep, giving the cardiovascular system a break from the demands of waking activity. Sleep disruption prevents this blood pressure dipping, contributing to hypertension. Adequate sleep is associated with lower risk of heart disease, stroke, and cardiovascular mortality, while chronic sleep deprivation increases these risks.

Growth and development are supported by sleep, particularly in children and adolescents. Growth hormone secretion peaks during slow-wave sleep, promoting physical growth and tissue repair. The large amounts of sleep typical of childhood and adolescence support the intense physical and neural development occurring during these periods.

From an evolutionary perspective, sleep likely evolved to serve multiple functions simultaneously, with natural selection favoring sleep because it conferred multiple benefits. The specific functions most important may vary across species depending on ecological pressures. For humans, memory consolidation, metabolic restoration, immune support, and emotional regulation all appear to be crucial functions that make sleep indispensable.

Dreams: The Sleeping Mind

Dreams are among the most fascinating and mysterious aspects of sleep. These vivid, often bizarre mental experiences that occur primarily during REM sleep have captivated human imagination for millennia, inspiring religious significance, artistic creation, and psychological theories. Modern neuroscience is beginning to understand the biology of dreaming, though many questions remain.

REM sleep dreams are typically vivid, visual, narrative experiences with clear story-like progression, emotional content, and often bizarre or impossible elements. The dreamer usually accepts the dream reality without question despite implausibilities that would be immediately recognized during waking. Dreams often involve the dreamer as a participant in the dream events rather than merely an observer. Memory for dreams is typically poor unless the person awakens directly from the dream, with most dreams being forgotten by morning.

Non-REM sleep also includes mental activity, though it is typically less vivid, less narrative, and more thought-like than REM dreams. Non-REM mentation may involve vague thoughts, fleeting images, or rumination about daily concerns. The most vivid non-REM dreams occur during stage 2 sleep, particularly later in the night when REM sleep pressure is high.

The neurobiology of REM dreaming reflects the unique brain state during REM sleep. The limbic system, including the amygdala, is highly active, potentially explaining the often intense emotional content of dreams. Visual and motor cortex are active, corresponding to the visual imagery and sense of movement in dreams. The prefrontal cortex, particularly dorsolateral regions involved in logic and executive control, shows reduced activity, potentially explaining why dreams often violate logical rules and why dreamers accept bizarre events without question.

The neurotransmitter environment during REM sleep, with high acetylcholine but very low norepinephrine and serotonin, may contribute to dream characteristics. Low norepinephrine may reduce the brain's ability to focus attention and maintain logical thought. Low serotonin may contribute to the often disorganized, illogical progression of dreams. High acetylcholine drives the activated cortical state supporting vivid imagery.

The activation-synthesis hypothesis, proposed by Allan Hobson and Robert McCarley, suggests that dreams result from the cortex trying to make sense of random neural activation generated by the brain stem during REM sleep. According to this theory, the pons generates random signals that activate visual, motor, and emotional regions. The cortex then attempts to synthesize these random signals into coherent narratives, creating the bizarre stories of dreams. This theory explains dream bizarreness as resulting from the cortex's attempt to integrate random, unrelated neural activity.

However, the activation-synthesis hypothesis has been criticized for portraying dreams as merely meaningless byproducts of random activity. Alternative theories suggest dreams serve important functions rather than being random noise. The threat simulation theory proposes that dreams evolved to allow rehearsal of threat responses in a safe context. The bias toward threatening and negative content in dreams may reflect this function, with the brain practicing responses to dangers.

The emotional regulation theory suggests dreams help process and integrate emotional experiences, particularly negative emotions. The high emotional content and low norepinephrine during REM sleep may allow reprocessing of emotional memories without triggering stress responses. This could explain why sleep helps people cope with stressful experiences and why REM sleep deprivation impairs emotional regulation.

The memory consolidation theory proposes that dreams reflect memory processing during sleep, with dream content representing memories being reviewed and integrated. The incorporation of recent experiences into dreams supports this view. However, most dream content is not straightforward replay of waking experiences but rather includes fragments, distortions, and combinations of multiple memories.

Dream content reflects both universal themes and individual experiences. Common dream themes across cultures include being chased, falling, being unprepared for tests or performances, losing teeth, and being unable to move or speak. These universal themes may reflect common concerns and evolutionary pressures. Individual dream content incorporates personal experiences, concerns, and memories, with recent experiences being particularly likely to appear in dreams, often in distorted forms.

The continuity hypothesis suggests that dream content continues waking concerns, emotions, and preoccupations. People dream about what occupies their waking minds, with dream emotions reflecting waking emotional states. Depression, anxiety, and trauma all affect dream content, with depressed individuals having more negative dream emotions, anxious individuals dreaming more about threats, and trauma victims experiencing recurring nightmares about traumatic events.

Lucid dreaming, the experience of being aware one is dreaming while the dream continues, has been studied in laboratory settings. Lucid dreamers can sometimes control dream content and signal to researchers using pre-arranged eye movements that can be detected on recordings. Brain activity during lucid dreams shows increased prefrontal cortex activation compared to non-lucid dreams, suggesting that the meta cognitive awareness of lucid dreaming involves regions suppressed during typical dreams.

The interpretation of dream meaning has ancient roots, with many cultures viewing dreams as messages from gods, spirits, or the unconscious mind. Sigmund Freud famously proposed that dreams represent disguised fulfillment of unconscious wishes, with manifest dream content concealing latent symbolic meanings requiring interpretation. However, modern neuroscience provides little support for the specific mechanisms Freud proposed, though the general idea that dreams reflect concerns and emotions remains reasonable.

Whether dreams have meaning or are merely random brain activity remains debated. The truth likely lies between these extremes. Dreams are not simply random but reflect brain systems activated during sleep, recent experiences, ongoing concerns, and emotional states. However, the meaning of dreams may not require deep interpretation but rather reflect relatively straightforward concerns and memories processed in the distinctive brain state of sleep.

Sleep Deprivation: The Consequences of Insufficient Sleep

Despite sleep's obvious importance, modern society systematically short-changes sleep. The average adult sleeps about 6.8 hours per night, below the 7 to 9 hours recommended for optimal health. Adolescents average even less relative to their needs, with school start times often conflicting with teenagers' naturally delayed sleep schedules. The consequences of this chronic sleep debt are profound and widespread.

Cognitive impairment from sleep deprivation is well-documented. Attention and vigilance decline markedly after even one night of partial sleep deprivation, making people prone to lapses in attention and errors. Working memory and executive functions including planning, problem-solving, and decision-making are impaired. Learning capacity declines, with sleep-deprived individuals encoding new information less effectively. Reaction time slows, increasing accident risk.

The magnitude of cognitive impairment from severe sleep deprivation equals or exceeds impairment from alcohol intoxication. Someone awake for 24 hours shows cognitive and motor impairment equivalent to a blood alcohol level of 0.10 percent, above the legal limit for driving in most jurisdictions. Yet while society takes drunk driving seriously, sleep-deprived driving receives far less attention despite causing thousands of deaths annually.

Micro sleeps, brief episodes of sleep lasting a few seconds, intrude during wakefulness when people are severely sleep-deprived. These micro sleeps can occur with eyes open, making them particularly dangerous during activities like driving. The person may not even realize they briefly lost consciousness. The monotony of highway driving is particularly conducive to micro sleeps, explaining why drowsy driving crashes often involve vehicles drifting off roads or into other lanes without any braking or evasive action.

Emotional regulation deteriorates with sleep deprivation, making people more irritable, moody, and prone to outbursts. The combination of heightened amygdala reactivity and reduced prefrontal control creates emotional volatility. Sleep-deprived individuals show biased processing of emotional information, attending more to negative than positive stimuli and interpreting ambiguous situations negatively. Social interactions become more difficult as sleep-deprived people struggle to read emotions accurately and respond appropriately.

Physical health consequences of chronic sleep deprivation are extensive. Metabolic regulation is disrupted, with sleep deprivation increasing appetite, particularly for high-calorie foods, while decreasing satiety signals. This combination promotes weight gain and obesity. Sleep deprivation also impairs glucose regulation and insulin sensitivity, increasing diabetes risk. Chronic short sleep is associated with increased risk of obesity, diabetes, cardiovascular disease, and all-cause mortality.

Immune function is compromised by inadequate sleep, increasing vulnerability to infections. People who sleep fewer than seven hours per night are three times more likely to develop colds when exposed to cold viruses compared to those sleeping eight or more hours. Chronic inflammation increases with sleep deprivation, contributing to numerous diseases. Wound healing slows in sleep-deprived individuals.

Mental health is profoundly affected by sleep. Insomnia is both a symptom of depression and a risk factor for developing depression. Sleep disruption can trigger manic episodes in bipolar disorder. Anxiety disorders are exacerbated by poor sleep. The relationship is bidirectional, with sleep problems contributing to mental illness and mental illness disrupting sleep, creating potential for vicious cycles.

Individual vulnerability to sleep deprivation varies, with some people showing substantial impairment after modest sleep restriction while others appear relatively resilient. Genetic factors contribute to this variation, with genes affecting circadian rhythms, adenosine signaling, and other sleep-related processes influencing how much sleep individuals need and how severely deprivation affects them. However, most people who believe they function well on minimal sleep show objective impairment they are unaware of, a phenomenon called sleep deprivation denial.

Recovery from sleep deprivation requires actually sleeping more, not simply resting or relaxing. One or two nights of extended sleep can reverse acute sleep deprivation effects, but chronic sleep debt accumulated over weeks or months may require extended recovery periods. Some effects of sleep deprivation, particularly if severe and prolonged, may not be fully reversible.

The causes of widespread sleep deprivation in modern society are multiple. Electric lighting allows extending wakefulness beyond natural dark periods, disrupting circadian rhythms. Work demands including long hours, shift work, and multiple jobs limit sleep time. Screen time before bed, particularly exposure to blue light from electronic devices, suppresses melatonin and delays sleep onset. Social pressures and cultural attitudes that valorize minimal sleep as productive and view adequate sleep as laziness contribute to sleep neglect. Stress and anxiety make falling asleep difficult. Underlying sleep disorders including insomnia and sleep apnea disrupt sleep quality.

Sleep Disorders: When Sleep Goes Wrong

Millions of people suffer from sleep disorders that prevent restorative sleep despite spending adequate time in bed. Understanding these disorders and their treatments is crucial for promoting health.

Insomnia, characterized by difficulty falling asleep, staying asleep, or early morning awakening despite adequate opportunity for sleep, is the most common sleep disorder, affecting roughly 10 to 15 percent of adults chronically and up to 30 percent occasionally. Insomnia causes daytime impairment including fatigue, mood disturbance, and cognitive impairment. It increases risk for depression, anxiety, substance abuse, and physical health problems.

Insomnia can be primary, occurring without other underlying conditions, or secondary to medical conditions, psychiatric disorders, medications, or other sleep disorders. However, this distinction has become less emphasized as research shows that even when insomnia initially develops secondary to other conditions, it often persists through conditioning and learned patterns even after the precipitating condition resolves.

The cognitive model of insomnia proposes that it develops and persists through a combination of predisposing factors including genetic vulnerability and personality traits, precipitating factors including stress or life changes that initially trigger sleep disruption, and perpetuating factors including maladaptive behaviors and thoughts about sleep. People with insomnia often develop anxiety about sleep itself, trying too hard to sleep, which paradoxically makes sleep more difficult. They may extend time in bed, nap during the day, or use alcohol to help sleep, behaviors that perpetuate insomnia.

Treatment for chronic insomnia typically begins with cognitive-behavioral therapy for insomnia (CBT-I), which targets perpetuating factors through multiple components. Sleep restriction limits time in bed to match actual sleep time, consolidating fragmented sleep. Stimulus control re associates the bed with sleep rather than wakefulness by restricting bed use to sleep and sex, leaving the bed if unable to sleep within 20 minutes. Cognitive therapy addresses anxious thoughts about sleep. Sleep hygiene education addresses behaviors that affect sleep. CBT-I is highly effective, producing lasting improvements in most patients without medication side effects or dependence risks.

Medications for insomnia including benzodiazepines, non-benzodiazepine hypnotics, and sedating antidepressants can provide short-term relief but are generally less effective than CBT-I for chronic insomnia and carry risks including tolerance, dependence, cognitive impairment, and rebound insomnia when discontinued. However, medications may be appropriate for short-term insomnia following acute stressors or as adjuncts to behavioral treatment.

Obstructive sleep apnea involves repeated interruptions of breathing during sleep due to collapse of the upper airway. These breathing pauses, which can occur hundreds of times per night, cause blood oxygen levels to drop and trigger brief arousals that fragment sleep. People with sleep apnea often snore loudly, gasp or choke during sleep, and wake unrefreshed despite spending adequate time in bed. Daytime sleepiness is prominent, creating safety risks particularly for driving.

Sleep apnea is associated with obesity, as excess tissue around the airway contributes to collapse. It increases risk for hypertension, heart disease, stroke, and diabetes through multiple mechanisms including intermittent hypoxia, sleep fragmentation, and activation of stress systems. Diagnosis requires overnight sleep study (polysomnography) documenting breathing disturbances.

Treatment typically involves continuous positive airway pressure (CPAP), where a mask worn during sleep delivers pressurized air that keeps the airway open. While highly effective when used consistently, many patients struggle with CPAP adherence due to discomfort or claustrophobia. Alternative treatments include oral appliances that reposition the jaw, positional therapy for patients whose apnea occurs primarily in certain positions, and surgical procedures to remove or reposition tissue, though surgery is typically reserved for cases where other treatments fail.

Narcolepsy involves excessive daytime sleepiness and sudden attacks of muscle weakness (cataplexy) triggered by strong emotions. It is caused by loss of orexin-producing neurons in the hypothalamus, as mentioned earlier. Without orexin's stabilizing influence on wake states, people with narcolepsy experience frequent intrusions of REM sleep features into wakefulness including cataplexy (loss of muscle tone like REM atonia), sleep paralysis (experiencing REM atonia while awake), and hypnagogic hallucinations (dream-like experiences at sleep onset).

Treatment combines medications to promote wakefulness during the day and sometimes medications to suppress REM sleep features. Scheduled naps can be helpful. Lifestyle modifications avoiding triggers for cataplexy and maintaining regular sleep schedules are important. While treatable, narcolepsy is a lifelong condition requiring ongoing management.

Restless legs syndrome involves uncomfortable sensations in the legs creating an irresistible urge to move them, particularly in the evening and at night. The sensations temporarily improve with movement but return at rest, making falling asleep difficult. Periodic limb movements during sleep, rhythmic leg movements that fragment sleep, often occur in conjunction with restless legs syndrome. Treatment may include iron supplementation if deficient, medications affecting dopamine signaling, and addressing underlying conditions that can trigger symptoms.

REM sleep behavior disorder, mentioned earlier, involves loss of the muscle paralysis that normally occurs during REM sleep, allowing people to physically act out dreams. This can result in injury to the person or bed partner. REM sleep behavior disorder most commonly affects older adults and is associated with neurodegenerative conditions including Parkinson's disease. Treatment may include medications to suppress REM sleep or increase muscle inhibition, and safety measures to prevent injury.

Sleep Across the Lifespan

Sleep needs and patterns change dramatically across the lifespan, reflecting developmental changes in brain structure and function.

Newborns sleep 14 to 17 hours per day in multiple short periods distributed across day and night. They spend roughly 50 percent of sleep time in REM sleep compared to 20 to 25 percent in adults, consistent with theories that REM sleep supports brain development. The circadian system is immature at birth, developing over the first months of life as infants gradually consolidate sleep into longer nighttime periods. Parents' understandable concern about infant sleep reflects the strain of disrupted sleep, though infant sleep patterns are developmentally appropriate.

Young children sleep 10 to 13 hours per night, with amounts gradually decreasing through childhood. Slow-wave sleep is abundant in childhood, supporting the intense physical and neural development occurring. Sleep problems in childhood including bedtime resistance, night wakings, and nightmares are common and usually resolve with consistent routines and parental reassurance.

Adolescents need 8 to 10 hours of sleep, though most get far less. A biological shift in circadian rhythms during puberty delays sleep onset, making teenagers naturally night owls. This creates conflict with early school start times, resulting in chronic sleep deprivation. Accumulating evidence shows that delaying school start times improves adolescent sleep, academic performance, and mental health, leading some school districts to implement later start times.

Adults need 7 to 9 hours of sleep, though individual variation exists. Sleep efficiency, the percentage of time in bed spent actually sleeping, is typically high in young adults but begins declining in middle age. Middle-aged adults often report more sleep disturbances including difficulty maintaining sleep and early morning awakening.

Older adults experience further changes in sleep architecture with reduced slow-wave sleep, more fragmented sleep, and earlier sleep and wake times reflecting circadian shifts. Sleep disorders including sleep apnea, restless legs syndrome, and insomnia become more common. Medical conditions and medications also affect sleep. However, the common belief that elderly people need less sleep is incorrect; they need similar amounts but have more difficulty obtaining good sleep. Addressing sleep problems in older adults can improve cognitive function, mood, and quality of life.

Conclusion: Prioritizing Sleep in Modern Life

Sleep is not a luxury or a waste of time but a biological necessity as important as nutrition and exercise for health and functioning. The brain and body cannot operate optimally without adequate sleep, and chronic sleep deprivation creates cascading problems across cognitive, emotional, and physical domains.

Modern life presents unprecedented challenges to healthy sleep through electric lighting, screens, work demands, and cultural attitudes that devalue sleep. Addressing the sleep crisis requires individual actions including prioritizing sleep, maintaining consistent schedules, creating sleep-conducive environments, and seeking treatment for sleep disorders, as well as societal changes including reasonable work hours, appropriate school start times, and cultural shifts recognizing sleep's importance.

Understanding the biology of sleep reveals why it cannot be shortchanged without consequences and provides guidance for optimizing sleep. The multistage architecture of sleep, the complex regulatory systems governing sleep and wake, the multiple functions sleep serves, and the profound impairments resulting from sleep deprivation all demonstrate that sleep is fundamental to brain health and optimal functioning.

The integration of sleep science into clinical practice, educational policy, and workplace culture has the potential to dramatically improve public health. Simple interventions including later school start times for adolescents, limiting shift work that requires sleeping during biological day, promoting sleep education, and treating sleep disorders could prevent countless accidents, improve mental health, enhance productivity, and extend healthy lifespan.

For individuals, the message is clear: sleep is an investment in cognitive performance, emotional well being, and physical health. The hours spent sleeping are not lost but rather enable better functioning during waking hours. Sacrificing sleep to gain productive time is counterproductive, as the cognitive and physical impairments from sleep deprivation undermine performance and health. Making sleep a priority, along with good nutrition and regular exercise, provides a foundation for optimal health and functioning.

13

DEPRESSION AND BRAIN CHEMISTRY

Depression is one of the most common and debilitating mental health conditions, affecting more than 300 million people worldwide. It is the leading cause of disability globally, surpassing cancer, heart disease, and diabetes in its impact on quality of life and functioning. Yet despite its prevalence and severity, depression remains widely misunderstood. It is often dismissed as mere sadness, weakness of character, or lack of willpower, leading to stigma that prevents many sufferers from seeking help.

Understanding depression as a biological brain disorder rather than a personal failing represents one of the most important advances in mental health. This shift in understanding has reduced stigma, improved treatments, and validated the experiences of millions of people whose suffering was previously dismissed or blamed on personal weakness. Depression is not simply feeling sad after disappointments or losses, though such normal reactions are sometimes confused with clinical depression. Rather, depression is a syndrome involving persistent changes in mood, cognition, motivation, and physical functioning that reflect alterations in brain chemistry and function.

The experience of depression extends far beyond sadness. People with major depression describe a profound emptiness, numbness, or despair that pervades all aspects of experience. Activities that once brought pleasure become empty and meaningless, a symptom called anhedonia that reflects dysfunction in the brain's reward systems. Concentration becomes difficult, with even simple tasks requiring enormous effort. Negative thoughts dominate, with people viewing themselves, their circumstances, and their futures through a lens of hopelessness. Sleep and appetite are typically disrupted, though in opposite directions for different people. Energy drains away, leaving profound fatigue unrelieved by rest. In severe cases, thoughts of death and suicide intrude, sometimes leading to tragic outcomes.

From a biological perspective, depression reflects disruptions in multiple brain systems and neurochemical pathways. While depression was once attributed to simple deficiencies in neurotransmitters like serotonin, current understanding recognizes that depression involves complex alterations in neural circuits, inflammatory processes, stress hormone regulation, and neural plasticity. This chapter explores the biology of depression, examining neurotransmitter systems, brain circuits, stress and inflammation, genetic and environmental risk factors, and how biological understanding informs treatment approaches that combine medication, psychotherapy, and lifestyle interventions.

Neurotransmitters and Depression: Beyond Simple Deficiency

The mono amine hypothesis of depression dominated thinking about depression's biology for decades. This hypothesis proposed that depression results from deficiencies in mono amine neurotransmitters including serotonin, norepinephrine, and dopamine. The theory emerged from observations that drugs depleting mono amines can cause depression, while drugs increasing mono amine levels often alleviate it. The development of selective serotonin re uptake inhibitors (SSRIs) and other antidepressants that increase mono amine levels seemed to confirm that depression involves mono amine deficiency.

However, the simple mono amine deficiency model has been challenged by several observations. Antidepressants increase mono amine levels within hours of first dose, yet therapeutic effects typically require weeks of treatment, suggesting the medications work through slower adaptive changes rather than simply correcting deficiencies. Some effective antidepressants do not primarily affect mono amine systems, while some drugs that increase mono amines are not effective antidepressants. People without depression given mono amine-enhancing drugs do not consistently experience mood improvements. These observations suggest that while mono amine systems are involved in depression, simple deficiency is not the complete explanation.

Serotonin appears particularly important for depression based on extensive evidence. Low levels of serotonin metabolites in cerebrospinal fluid are associated with depression, particularly depression accompanied by suicidality and impulsive aggression. Depleting tryptophan, the amino acid precursor to serotonin, can temporarily worsen mood in people with history of depression or family history of depression, though it typically does not affect mood in healthy individuals. Genetic variations affecting serotonin function are associated with increased depression risk, particularly when combined with stressful life experiences.

SSRIs including fluoxetine, sertraline, paroxetine, citalopram, and escitalopram work by blocking the serotonin transporter, the protein that removes serotonin from synapses back into presynaptic neurons. By blocking reuptake, SSRIs increase serotonin levels in synapses, enhancing serotonin signaling. These medications are effective for many but not all depressed patients, with roughly 50 to 65 percent showing significant improvement compared to about 30 to 45 percent responding to placebo. This means SSRIs help substantially more people than placebo, but responses are incomplete for many and some people do not benefit at all.

Serotonin-norepinephrine reuptake inhibitors (SNRIs) including venlafaxine, duloxetine, and desvenlafaxine block reuptake of both serotonin and norepinephrine. These medications are similarly effective to SSRIs overall, though some patients respond better to one class than the other. Norepinephrine appears particularly important for energy, motivation, and concentration, potentially explaining why SNRIs may be more effective for depression characterized by fatigue and poor concentration.

Norepinephrine's role in depression is supported by effectiveness of older tricyclic antidepressants and monoamine oxidase inhibitors that enhance norepinephrine function. Brain imaging studies show altered norepinephrine system function in depression. The locus coeruleus, the brain's primary source of norepinephrine, shows altered activity in depression. Chronic stress, a major risk factor for depression, dysregulates norepinephrine systems.

Dopamine, the neurotransmitter central to motivation and reward, appears crucial for anhedonia, the inability to experience pleasure that is a core feature of depression. Reduced dopamine signaling in reward circuits including the ventral striatum is associated with anhedonia. Brain imaging studies show that depressed individuals exhibit blunted neural responses to rewards. Medications that enhance dopamine function including bupropion, which blocks dopamine and norepinephrine reuptake, can be effective antidepressants, particularly for depression with prominent anhedonia.

Beyond the mono amines, other neurotransmitter systems contribute to depression. Glutamate, the brain's primary excitatory neurotransmitter, shows altered function in depression, with some evidence of excessive glutamate activity in certain brain regions. Ketamine, an anesthetic that blocks glutamate NMDA receptors, produces rapid antidepressant effects in many treatment-resistant patients, often within hours rather than the weeks required for traditional antidepressants. This dramatic response has generated enormous interest in glutamate-based treatments, though ketamine's effects are often temporary and questions remain about optimal use.

GABA, the primary inhibitory neurotransmitter, also appears altered in depression, with reduced GABA levels in certain brain regions. This might contribute to excessive rumination and negative thinking if inhibitory control of repetitive thought patterns is impaired.

The role of neurotransmitters in depression is clearly complex, involving not simple deficiencies but rather imbalances and dysregulation across multiple systems. Antidepressants likely work not by simply restoring neurotransmitter levels but by triggering adaptive changes in how neurons respond to neurotransmitters, in receptor sensitivity and number, in intracellular signaling cascades, and ultimately in gene expression and neural plasticity. These downstream effects develop over weeks, explaining the delayed therapeutic response.

Neural Circuits and Brain Structure in Depression

Depression involves dysfunction in specific neural circuits that regulate mood, motivation, and cognitive processes. Brain imaging has revealed structural and functional abnormalities in multiple brain regions in depression, providing insight into how depression disrupts normal brain function.

The prefrontal cortex, particularly the dorsolateral and ventromedial regions, shows reduced activity and often reduced volume in depression. These regions support executive functions including planning, decision-making, working memory, and emotion regulation. Reduced prefrontal function may contribute to the cognitive impairments, poor concentration, and difficulty regulating negative emotions characteristic of depression.

The anterior cingulate cortex, which detects conflicts, signals the need for cognitive control, and processes emotional aspects of pain, shows altered function in depression. Some studies find hyperactivity while others find hypoactivity, possibly reflecting different depression sub types or states. Pregenual anterior cingulate cortex appears particularly important for depression, with this region showing reduced volume and altered connectivity in depressed individuals.

The amygdala, central to processing emotions particularly negative emotions, often shows hyperactivity in depression. This hyperactivity may contribute to the bias toward negative emotional processing characteristic of depression, where depressed individuals selectively attend to negative information, interpret ambiguous situations negatively, and recall negative memories more readily than positive ones. The amygdala's heightened reactivity to negative stimuli combined with reduced prefrontal regulation creates a brain stuck in negative emotional states.

The hippocampus, crucial for memory and stress regulation, consistently shows reduced volume in depression, with the degree of reduction correlating with duration of untreated depression. Multiple factors contribute to hippocampal atrophy including chronic stress and cortisol elevation, which can damage hippocampal neurons and suppress neurogenesis. Reduced hippocampal volume may impair contextual memory and contribute to overgeneralized negative memories where specific negative experiences are recalled as representative of life in general.

The ventral striatum, including the nucleus accumbens that is central to reward processing, shows blunted responses to rewards in depression. This dysfunction in reward circuitry provides a neural correlate for anhedonia. Brain imaging studies show that when depressed individuals view or anticipate rewards, the ventral striatum activation is markedly reduced compared to non-depressed individuals. Successful antidepressant treatment often normalizes these reward responses.

The subgenual cingulate cortex, also called Brodmann area 25, has emerged as particularly important in depression. This region shows elevated activity in depression that decreases with successful treatment regardless of whether treatment is medication or psychotherapy. Deep brain stimulation targeting this region has shown promise for treatment-resistant depression, though the approach remains experimental.

Connectivity between these regions appears disrupted in depression. The default mode network, a set of brain regions active during rest and self-referential thought, shows altered connectivity. This may contribute to excessive rumination, the repetitive negative thinking characteristic of depression where people obsessively replay problems and negative experiences without productive problem-solving.

White matter abnormalities affecting connections between brain regions are increasingly recognized in depression. Diffusion tensor imaging studies show reduced white matter integrity in tracts connecting frontal and limbic regions. These connectivity disruptions may impair communication between brain regions involved in cognitive control and emotional processing, reducing the ability to regulate mood through cognitive strategies.

These structural and functional brain changes in depression are not simply consequences of being depressed but may also create vulnerability to depression and contribute to its persistence. Some brain changes appear to normalize with successful treatment, suggesting they are state markers that track with current depression severity. Others persist even after recovery and are also seen in never-depressed individuals at high risk for depression, suggesting they are trait markers reflecting vulnerability.

The observation that brain structure and function are altered in depression has important implications. It validates depression as a genuine medical condition involving measurable brain abnormalities rather than a character flaw or weakness. It suggests that effective treatment may need to address these neural circuit dysfunctions, whether through medications that alter neurotransmitter function, psychotherapy that engages and potentially reorganizes neural circuits, or other interventions targeting brain function.

Stress, Inflammation, and Depression

The relationship between stress and depression is well-established, with stressful life events being among the strongest predictors of depression onset. Understanding the biological mechanisms linking stress to depression has revealed that chronic stress produces changes in the brain and body that create vulnerability to depression.

The hypothalamic-pituitary-adrenal (HPA) axis, the body's primary stress response system discussed in earlier chapters, shows profound dysregulation in many depressed individuals. Cortisol, the stress hormone released by the adrenal cortex, is often elevated in depression. This elevation appears to reflect impaired negative feedback, where cortisol fails to properly suppress its own production by inhibiting the hypothalamus and pituitary. As a result, the HPA axis remains activated even in the absence of stress.

Chronic cortisol elevation has multiple harmful effects relevant to depression. As discussed in Chapter 7, sustained high cortisol damages the hippocampus, reducing its volume and impairing its functions in memory and mood regulation. Cortisol also affects prefrontal cortex function, potentially impairing cognitive control and decision-making. Cortisol influences neurotransmitter systems, altering serotonin and dopamine function in ways that could contribute to depression.

Not all depressed individuals show elevated cortisol; some show blunted cortisol responses, and the pattern may vary across depression subtypes. However, HPA axis dysregulation of some form appears common in depression. Importantly, HPA axis function often normalizes with successful depression treatment, suggesting the dysregulation is related to depression rather than being a separate problem.

The role of inflammation in depression has become a major focus of research in recent years. Multiple lines of evidence support an inflammatory component to depression. Depressed individuals show elevated levels of inflammatory cytokines including interleukin-6, tumor necrosis factor-alpha, and C-reactive protein. These inflammatory markers correlate with depression severity and predict treatment response.

Inflammatory cytokines can directly affect the brain, crossing the blood-brain barrier or signaling through neural pathways from the periphery to brain. Once in the brain, inflammatory signals affect neurotransmitter metabolism, reduce neuroplasticity, alter neural circuit function, and can produce the constellation of symptoms called sickness behavior that resembles depression including fatigue, anhedonia, social withdrawal, and sleep disturbance.

The sources of inflammation in depression are multiple. Chronic stress activates inflammatory pathways. Obesity is associated with chronic inflammation and also with increased depression risk. Autoimmune conditions and other inflammatory diseases show high rates of depression. Early life stress and trauma may program inflammatory systems to be hyperreactive. Genetic variations affecting immune function influence depression risk.

This inflammatory perspective on depression has treatment implications. Anti-inflammatory drugs show antidepressant effects in some studies, though results are mixed. Lifestyle interventions including exercise, healthy diet, and stress reduction that reduce inflammation may help prevent or treat depression partly through anti-inflammatory effects. Understanding that some depression may have an inflammatory basis could lead to targeted treatments and better matching of patients to treatments based on their inflammatory profiles.

The gut-brain axis represents another pathway through which stress and inflammation may contribute to depression. The gut microbiome, the trillions of bacteria living in the intestines, communicate with the brain through multiple pathways including the vagus nerve, immune signaling, and production of neurotransmitter precursors. Stress alters the gut microbiome composition, and alterations in microbiome composition are associated with depression. Animal studies show that transferring gut microbiota from depressed individuals to germ-free mice can induce depression-like behaviors.

Whether microbiome changes contribute to depression or are consequences of depression-related changes in diet, stress, and behavior remains unclear. However, the possibility that modifying the microbiome through probiotics, prebiotics, or diet might affect mood has generated considerable interest. While evidence remains preliminary, this represents another pathway through which biological factors beyond the brain proper might influence mental health.

Genetics and Environment in Depression

Depression runs in families, with first-degree relatives of depressed individuals showing two to three times higher risk than the general population. Twin studies estimate heritability of depression at 30 to 40 percent, meaning genetic differences account for roughly one-third of variation in depression risk while environmental factors account for the remainder. This moderate heritability indicates that both genetic vulnerability and environmental triggers contribute to depression.

Molecular genetics has identified numerous genetic variants associated with depression risk, though each individual variant has very small effects. The largest genome-wide association studies have identified hundreds of genetic loci associated with depression, collectively explaining about 10 percent of variance in depression risk. These genes affect diverse biological processes including neurotransmitter function, neural development, stress response systems, and immune function.

One of the most studied genetic variants is in the serotonin transporter gene (5-HTTLPR). The short allele of this polymorphism is associated with reduced serotonin transporter expression and has been reported to increase depression risk, particularly in combination with stressful life events. However, this gene-by-environment interaction has been controversial, with some studies replicating it and others failing to find effects. Meta-analyses suggest the effect is real but small, and the gene alone has minimal impact without environmental stress.

This gene-by-environment interaction illustrates an important principle: genetic risk for depression is not deterministic but rather creates vulnerability that may or may not lead to depression depending on environmental factors. People carrying higher genetic risk are more vulnerable to depression following stress, trauma, or other adversity, while those with lower genetic risk show more resilience. Conversely, genetic risk matters less in environments with minimal stress and good support.

The diathesis-stress model captures this interaction, proposing that depression results from the combination of diathesis (predisposition or vulnerability) and stress (environmental triggers). Neither alone is sufficient; depression develops when vulnerable individuals encounter sufficient stress. This model helps explain why depression develops in some people but not others following similar stressors, and why some highly vulnerable individuals remain healthy in supportive environments.

Environmental risk factors for depression are numerous and well-documented. Stressful life events including loss, trauma, chronic difficulties, and major changes predict depression onset, particularly in vulnerable individuals. Childhood adversity including abuse, neglect, parental loss, and household dysfunction creates lasting vulnerability to adult depression through mechanisms including altered stress response systems, learned helplessness, and disrupted attachment patterns.

Social isolation and lack of social support increase depression risk, while strong social connections protect against depression even in the face of stress. The quality of close relationships appears particularly important, with conflictual or unsupportive relationships increasing risk while warm, supportive relationships providing protection.

Certain medical conditions increase depression risk, including thyroid disorders, vitamin D deficiency, chronic pain, cardiovascular disease, diabetes, and neurodegenerative diseases. Some medications can trigger depression as a side effect. Substance abuse, particularly alcohol abuse, both results from and contributes to depression in bidirectional relationships.

Seasonal patterns affect mood in some individuals, with seasonal affective disorder (SAD) involving recurrent depression during winter months when daylight is reduced. Light exposure appears crucial, with reduced sunlight affecting circadian rhythms, melatonin production, and mood. Light therapy using bright light boxes can effectively treat SAD, demonstrating an environmental intervention targeting biological mechanisms.

Understanding the interplay of genetic and environmental factors in depression has several implications. It suggests that preventing depression requires addressing environmental risk factors including childhood adversity, chronic stress, and social isolation, while also identifying and supporting vulnerable individuals. It implies that effective treatment may need to address both biological vulnerabilities through medication or other biological interventions, and environmental and psychological factors through psychotherapy and social support.

Treatment: Biological and Psychological Approaches

Treatment for depression has advanced considerably based on growing understanding of its biological foundations, though significant room for improvement remains. Effective treatments include antidepressant medications, psychotherapy, brain stimulation techniques, and lifestyle interventions, often used in combination.

Antidepressant medications remain the most common biological treatment. As discussed, these medications primarily work by increasing monoamine neurotransmitter levels, though their therapeutic effects likely reflect downstream adaptive changes rather than simply correcting deficiencies. Modern antidepressants including SSRIs and SNRIs are generally safer and better tolerated than older medications, though all carry potential side effects and not all patients respond.

The time course of antidepressant response is important for managing expectations. Therapeutic effects typically begin emerging after two to four weeks of treatment, with full effects sometimes requiring eight to twelve weeks. Side effects often appear immediately, before benefits emerge, leading some patients to discontinue prematurely. Understanding that delayed benefit is expected helps patients persist through initial weeks.

Treatment-resistant depression, where patients do not respond adequately to multiple antidepressant trials, affects roughly one-third of patients. For these individuals, alternative approaches include switching medication classes, combining medications, augmentation strategies adding other medications to boost antidepressant effects, or trying novel treatments.

Ketamine, mentioned earlier, represents one novel approach showing dramatic effects in treatment-resistant depression. Unlike traditional antidepressants requiring weeks for effect, ketamine often produces improvements within hours that can last days to weeks. The mechanisms remain under investigation but appear to involve rapid increases in glutamate signaling and enhanced neural plasticity. Esketamine, a ketamine derivative delivered as a nasal spray, has been approved for treatment-resistant depression, though concerns about long-term effects and optimal usage patterns remain.

Psychedelics including psilocybin (from "magic mushrooms") are being investigated as antidepressants based on preliminary evidence of rapid and sustained improvements following one or two doses combined with psychotherapy. These substances appear to promote neural plasticity and produce profound psychological experiences that patients describe as therapeutic. However, research is still in early stages, and clinical use remains experimental.

Electroconvulsive therapy (ECT), despite its frightening reputation from misuse in earlier eras, remains the most effective treatment for severe, treatment-resistant depression. Modern ECT is safe and humane, performed under anesthesia with muscle relaxants to prevent injury. It produces seizures that somehow reset dysfunctional neural circuits, though precise mechanisms remain unclear. ECT is particularly effective for severe depression with psychotic features, catatonia, or high suicide risk, though memory side effects limit its use.

Transcranial magnetic stimulation (TMS) uses magnetic fields to stimulate specific brain regions non-invasively. Repeated sessions of TMS targeting the dorsolateral prefrontal cortex show antidepressant effects with minimal side effects. TMS is approved for treatment-resistant depression and represents a middle ground between medication and ECT in terms of efficacy and invasiveness.

Psychotherapy, particularly cognitive-behavioral therapy (CBT) and interpersonal therapy, is highly effective for depression, with effects comparable to medication for mild to moderate depression. CBT helps patients identify and modify negative thought patterns, develop more adaptive beliefs, and change behaviors that maintain depression. Interpersonal therapy focuses on improving relationship patterns and resolving interpersonal problems that contribute to depression.

Brain imaging studies show that psychotherapy produces measurable changes in brain function similar to medication, including normalized activity in prefrontal cortex, anterior cingulate, and other regions. This provides biological validation that psychotherapy is not just "talk" but produces genuine changes in brain function underlying mood improvement.

Combining medication and psychotherapy often produces better outcomes than either alone, particularly for moderate to severe depression or chronic/recurrent depression. Medication may provide initial symptom relief allowing patients to engage in therapy, while therapy provides skills and insights that prevent relapse after medication discontinuation.

Lifestyle interventions including exercise, sleep improvement, stress management, and social connection provide important adjuncts to treatment. Exercise shows antidepressant effects comparable to medication for mild to moderate depression, likely through multiple mechanisms including increased neurotrophic factors, reduced inflammation, improved sleep, and behavioral activation. Even modest amounts of aerobic exercise or strength training can benefit mood.

Sleep disturbance both contributes to and results from depression. Addressing sleep through behavioral interventions or medication can improve depression outcomes. Exposure to bright light, particularly morning light, helps regulate circadian rhythms and can improve mood, especially in seasonal depression.

Social connection and behavioral activation, engaging in meaningful activities and relationships despite not feeling motivated, help break cycles of withdrawal and inactivity that maintain depression. While depressed individuals often isolate themselves, forcing engagement through structured activities and social contact can gradually improve mood.

Prevention and Long-term Management

Given the recurring nature of depression, with 50 to 80 percent of people experiencing additional episodes after their first, prevention of recurrence is crucial. Continuation treatment with medication or maintenance psychotherapy after recovery reduces relapse risk substantially. Patients who stop treatment too early show high relapse rates, while those continuing for at least six to twelve months after recovery are much more likely to maintain improvement.

For people with multiple previous episodes, longer-term or even indefinite maintenance treatment may be warranted. While this represents significant burden and potential for ongoing side effects, the alternative of repeated severe depressive episodes often represents greater burden and risk.

Beyond medical treatment, understanding personal vulnerability factors and early warning signs allows individuals to seek help early when symptoms return. Developing robust coping strategies, maintaining social connections, managing stress effectively, and addressing problems before they accumulate all help prevent recurrences.

At a societal level, preventing depression requires addressing environmental risk factors including childhood adversity, chronic stress, poverty, social isolation, and lack of access to healthcare. Early intervention programs supporting vulnerable families, school-based programs teaching coping skills and promoting mental health, workplace policies reducing chronic stress, and anti-stigma campaigns encouraging help-seeking all contribute to reducing depression's burden.

Conclusion: Depression as a Brain Disorder

The accumulated evidence that depression involves measurable alterations in brain chemistry, structure, function, and connectivity firmly establishes it as a biological brain disorder, not a character flaw or personal weakness. This understanding reduces stigma, validates sufferers' experiences, and points toward treatments targeting biological mechanisms.

However, recognizing depression's biological foundations does not reduce it to purely biological phenomena. Depression emerges from complex interactions between genetic vulnerability, brain chemistry, neural circuits, stress exposure, inflammatory processes, psychological patterns, social relationships, and life circumstances. Effective treatment and prevention must address multiple levels simultaneously, combining biological interventions that alter brain function with psychological and social interventions that address thoughts, behaviors, relationships, and environments.

The brain's remarkable plasticity, discussed in earlier chapters, provides hope that even longstanding depression can improve through interventions that promote adaptive neural changes. Whether through medication triggering molecular cascades that enhance plasticity, psychotherapy engaging circuits in new patterns, lifestyle changes reducing inflammation and stress, or combinations of approaches, recovery involves the brain reorganizing itself in healthier patterns.

In the final chapter, we will synthesize insights from throughout this book to reflect on the remarkable organ that is the human brain, its capacities and vulnerabilities, how biology and experience intertwine to create who we are, and how understanding biological psychology empowers us to protect brain health, maximize human potential, and approach mental health challenges with compassion and science-based interventions.

14

THE FUTURE OF BIOLOGICAL PSYCHOLOGY

We have journeyed through the biological foundations of human psychology, exploring the brain structures that create thought and emotion, the neurons and neurotransmitters that transmit information, the hormones that shape behavior, the plasticity that allows learning, and the disorders that emerge when these systems malfunction. This exploration reveals a profound truth: we are biological beings, and everything we experience as mental life every thought, feeling, memory, and dream emerges from the physical activity of our brains.

Yet this biological perspective does not reduce human experience to mere mechanism or determinism. Rather, it reveals the brain as a remarkably sophisticated organ that transforms physical and chemical processes into the richness of human consciousness. It shows how genes and environment intertwine in complex ways, how early experiences shape neural development, how adult brains remain plastic and capable of change, and how understanding biology empowers us to intervene when things go wrong.

The field of biological psychology has made extraordinary progress over the past century. We have mapped the brain's major structures and identified their functions. We have discovered neurotransmitters and how they work. We have observed living brains in action through imaging technologies. We have identified genes influencing behavior and mental illness. We have developed medications and therapies that can alleviate suffering. We have documented the brain's capacity for reorganization and recovery.

Yet for all this progress, much remains unknown. We still do not fully understand how neural activity creates subjective conscious experience, why some people develop mental illness while others remain resilient, how to effectively treat many brain disorders, or how to unlock the brain's full potential for learning and adaptation. The coming decades promise continued discoveries that will deepen understanding and improve interventions.

This final chapter explores where biological psychology is heading, examining emerging technologies and approaches that are transforming the field, unresolved questions that drive continued research, ethical challenges raised by growing ability to measure and modify brain function, and how biological understanding can be applied to promote human flourishing while avoiding misuse. We will synthesize insights from throughout the book to reflect on what biological psychology teaches us about human nature and how this knowledge can guide individual choices and societal policies.

Emerging Technologies: New Windows into the Brain

The history of biological psychology is inseparable from the history of technology. Each major advance in our ability to observe and measure brain function has revolutionized understanding. Electroencephalography revealed that sleep involves distinct stages. Brain imaging showed which regions activate during different mental tasks. Optogenetics allowed precise control of specific neurons in animals. New technologies on the horizon promise to continue this pattern of technology-driven discovery.

Advanced brain imaging techniques are providing unprecedented detail about brain structure and function. High-field MRI scanners with magnetic fields of 7 Tesla or stronger produce images with sub millimeter resolution, revealing fine structures invisible to conventional scanners. These ultra-high-resolution images allow detailed mapping of cortical layers, white matter tracts, and small sub cortical structures, providing anatomical precision approaching that of postmortem examination but in living brains.

Functional imaging is becoming faster and more precise. Traditional fMRI has temporal resolution of seconds, too slow to capture the rapid dynamics of neural processing. New techniques including magnetoencephalograp hy (MEG) and high-density EEG provide millisecond temporal resolution while advances in analysis methods improve spatial localization. Combining multiple imaging modalities provides both excellent spatial and temporal resolution, tracking neural activity across the brain with unprecedented precision.

Molecular imaging using PET scans with targeted radioligands allows visualization of specific neurotransmitter systems, receptor densities, and metabolic processes. New radioligands are being developed to image inflammation, protein aggregation in neurodegenerative diseases, and other molecular processes relevant to brain function and disease. This allows researchers to observe neurochemistry in living humans, connecting molecular processes to symptoms and treatment responses.

Optogenetics, which uses light to control genetically modified neurons expressing light-sensitive proteins, has revolutionized animal neuroscience by allowing researchers to activate or silence specific neuron populations with millisecond precision. While optogenetics cannot currently be used in humans due to requirements for genetic modification and surgical implantation of light sources, it has provided crucial insights into how specific neural circuits control behavior. Future developments might allow less invasive optogenetic approaches or analogous techniques applicable to humans.

Chemogenetics, similar to optogenetics but using designer drugs to activate designer receptors expressed in specific neurons, provides another tool for controlling neural activity with high specificity. DREADD (Designer Receptors Exclusively Activated by Designer Drugs) technology allows researchers to activate or inhibit specific neuron populations by administering drugs that only affect cells expressing the designer receptors. This approach is less invasive than optogenetics and may have more near-term transnational potential.

Brain-computer interfaces that decode neural signals and use them to control devices are advancing rapidly. Paralyzed individuals have learned to control robotic arms, computer cursors, and communication devices using signals recorded from motor cortex. As decoding algorithms improve and recording technology advances, brain-computer interfaces may provide increasingly sophisticated ways for people to interact with technology and for researchers to understand how the brain encodes information.

Closed-loop systems that both record from and stimulate the brain based on detected patterns represent the next generation of neural interfaces. These systems can detect abnormal activity like seizures and deliver stimulation to interrupt them, or detect mood states and provide targeted therapy. Deep brain stimulation for Parkinson's disease is evolving toward adaptive systems that adjust stimulation based on real-time monitoring of neural activity, potentially improving efficacy while reducing side effects.

Artificial intelligence and machine learning are transforming how neuroscience data are analyzed. Deep learning algorithms can identify patterns in brain imaging data that predict diagnoses, treatment responses, or future outcomes. Machine learning can decode mental states from brain activity patterns, potentially allowing objective measurement of subjective experiences. AI can also generate hypotheses by identifying correlations in large datasets that human researchers might miss.

However, these powerful analytical tools also raise concerns. Machine learning models often work as "black boxes," making predictions without providing interpretable explanations of how they reached conclusions. This can be problematic when predictions guide clinical decisions. Ensuring AI tools are reliable, unbiased, and used appropriately requires careful validation and ongoing oversight.

Organoid technology, where brain-like structures are grown from stem cells in laboratory dishes, provides new models for studying human brain development and disease. Brain organoids develop some of the layered structure and cell types of real brains and can be generated from patients' own cells, allowing study of how genetic variations affect brain development. While current organoids are far simpler than real brains and lack many features including blood vessels and connectivity, they provide unprecedented opportunities to study human neural development and test potential therapies.

Gene editing technologies including CRISPR allow precise modification of genetic sequences, opening possibilities for correcting genetic mutations that cause brain disorders. While application in humans remains largely experimental due to ethical and safety concerns, gene editing has enormous potential for treating genetic forms of intellectual disability, neurodegenerative disease, and other conditions. The ethical implications of human genetic modification, particularly heritable germline editing, require careful societal deliberation.

Unresolved Questions: The Frontiers of Understanding

Despite remarkable progress, fundamental questions about how the brain creates mind remain unanswered. These questions drive ongoing research and will likely occupy neuroscientists for decades to come.

The hard problem of consciousness, as philosopher David Chalmers termed it, asks how and why physical processes in the brain give rise to subjective experience. Why does seeing red feel like something rather than being merely information processing? How does the brain bind together distributed neural activity into unified conscious experiences? Neuroscience has made progress on the "easy problems" of consciousness, identifying neural correlates of specific conscious contents and states. But explaining why these neural processes feel like something, why there is "something it is like" to be a conscious being, remains deeply mysterious.

Some theorists propose that consciousness emerges from specific types of information integration, while others suggest it requires particular patterns of neural synchrony or feedback loops. Integrated Information Theory proposes that consciousness arises from systems that integrate information in specific ways, quantifiable through measures of integrated information. Global Workspace Theory suggests consciousness arises when information is broadcast widely across the brain through a global workspace. These theories make testable predictions, but consciousness remains one of neuroscience's greatest mysteries.

The nature of memory storage is incompletely understood despite extensive research on memory mechanisms. We know that memory involves changes in synaptic strengths and that specific brain regions support different types of memory. But precisely how information is encoded in patterns of synaptic weights, how these patterns are maintained over decades, and how the brain retrieves specific memories from vast stores of information remain unclear. Understanding memory encoding at this detailed level could reveal why memories sometimes fail and how to enhance memory function.

The genetic architecture of complex traits including intelligence, personality, and mental illness involves thousands of genetic variants each with tiny effects. Understanding how these variants collectively influence brain development and function, how they interact with each other and with environmental factors, and whether this knowledge can guide personalized interventions requires integrating genetics, neuroscience, and sophisticated computational modeling. Current genetic findings explain only small fractions of heritability, suggesting much remains to be discovered about genetic influences on brain and behavior.

Individual differences in brain organization and function are enormous, yet most neuroscience research has focused on finding commonalities across individuals. Understanding how brains differ and what these differences mean for cognition, emotion, and behavior requires large-scale studies tracking many individuals over time. The growing availability of large datasets from projects including the Human Connectome Project and UK Biobank is enabling research on individual differences, but linking brain variations to meaningful psychological differences remains challenging.

The relationship between neural activity and behavior is often studied as if neural activity causes behavior, but the relationship is bidirectional. Behavior and experience modify neural activity and brain structure through plasticity. Teasing apart these bidirectional influences requires experimental manipulations or sophisticated statistical approaches accounting for reciprocal causation. Understanding these feedback loops is crucial for interventions, as changing behavior can change the brain just as changing the brain can change behavior.

Mental illness classification remains largely based on symptom clusters rather than biological mechanisms. Different patients receiving the same diagnosis may have different underlying biological abnormalities requiring different treatments. Research Domain Criteria (RDoC), an initiative to reclassify mental illness based on neuroscience and biology rather than symptoms alone, aims to identify biologically coherent subgroups within current diagnostic categories. This could enable more precise matching of patients to treatments, but requires validating biological markers and demonstrating they predict treatment response.

Ethical Considerations: Power and Responsibility

As biological psychology advances, it raises profound ethical questions about how knowledge should be used and what limits should be placed on applications. The ability to measure, predict, and potentially modify brain function creates both opportunities and risks that society must navigate carefully.

Neuroimaging in legal contexts has generated controversy. Brain imaging showing abnormalities in defendants' brains has been used in criminal trials to argue for reduced responsibility or to predict future dangerousness. However, the validity of using brain scans for these purposes is questionable. Brain differences observed at the group level may not meaningfully predict individual behavior. The logic that brain abnormalities excuse criminal behavior could paradoxically lead to longer sentences if abnormalities are interpreted as indicating high risk of reoffending.

The legal system's concepts of free will, responsibility, and culpability may need to evolve as neuroscience reveals more about how brains make decisions. If all decisions emerge from prior brain states determined by genes and experience, does the concept of free will remain coherent? Most neuroscientists and philosophers argue that neuroscience findings are compatible with holding people responsible for their actions, but ongoing dialogue between neuroscience and law is needed to ensure legal principles remain grounded in scientific understanding while preserving justice.

Cognitive enhancement through drugs, brain stimulation, or other technologies raises questions about fairness, coercion, and human nature. If safe and effective cognitive enhancement becomes available, should everyone have access, or would costs create new inequalities? Would people feel pressured to enhance to remain competitive? Would enhancement fundamentally alter what it means to be human? These questions parallel ongoing debates about genetic enhancement but add the dimension of directly modifying brain function.

Some cognitive enhancement already occurs through caffeine, prescription stimulants used off-label by students and professionals, and other means. As enhancement technologies improve, societies will need to decide what enhancements are acceptable, who should have access, and what regulations are needed. Blanket prohibition seems unlikely to succeed and may be undesirable if enhancements improve wellbeing without unacceptable costs. But unrestricted access could lead to coercive enhancement pressures and exacerbate inequalities.

Privacy of mental states becomes a concern as brain reading technologies improve. If patterns of brain activity can reveal thoughts, emotions, intentions, or deception, should this information be accessible to employers, governments, or others? Current technology cannot read specific thoughts, but it can decode general mental states, detect recognition of familiar stimuli, and predict some decisions before conscious awareness. As decoding improves, protecting neural privacy may require legal frameworks similar to those protecting genetic information.

Neurotechnology implants including brain-computer interfaces raise questions about identity, autonomy, and security. If neural implants enhance cognitive abilities or provide new capabilities, how do they affect personal identity? If implants are hacked or malfunction, could they harm users or compromise their agency? Who owns data generated by neural implants? These questions will become more pressing as neural implants transition from purely medical applications to potential enhancement uses.

Informed consent in neuroscience research and clinical care requires ensuring people understand what brain interventions entail, including both potential benefits and risks. Some interventions like deep brain stimulation can have dramatic effects on personality and emotion, raising questions about authentic consent when the intervention might alter the very self making the decision. Research involving vulnerable populations including children, people with dementia, or prisoners requires special protections.

Stigma reduction is both an opportunity and a risk from biological psychiatry. Understanding mental illness as brain disorders reduces blame and shame, but it can also promote biological essentialism where people are reduced to their diagnoses. Emphasizing that mental illness has biological underpinnings helps combat the view that it reflects personal weakness, but it must be balanced with recognition that people are not defined by diagnoses and that recovery is possible.

Applications: From Understanding to Action

Biological psychology provides more than theoretical understanding; it offers practical applications for improving lives. Translating research findings into effective interventions remains a major focus as the field matures.

Precision medicine in psychiatry aims to match individual patients to optimal treatments based on their biological profiles. Rather than trial-and-error approaches where patients try multiple medications before finding one that works, precision approaches would use genetic testing, brain imaging, or other biomarkers to predict treatment response. Progress has been limited, with few biomarkers reliably predicting treatment response, but ongoing research continues pursuing this goal.

Pharmacogenomics, which studies how genetic variations affect drug responses, has identified variants affecting metabolism of some psychiatric medications, allowing dose adjustments to avoid side effects or therapeutic failures. As understanding of genetic influences on drug response improves, genetic testing may increasingly guide medication selection, though most treatment response variation remains unexplained by currently known genetic variants.

Early intervention for mental illness based on identification of prodromal symptoms or biological markers could prevent full-blown disorders from developing. Adolescents showing early signs of psychosis can be offered interventions that may prevent or delay schizophrenia onset. Children with high genetic risk for depression can be monitored and offered preventive interventions if early symptoms appear. However, early intervention raises ethical concerns about false positives, stigmatization, and treating people who might never have developed illness.

Digital mental health interventions including smartphone apps, online therapy, and computer-based cognitive training provide scalable approaches to delivering mental health care. These technologies allow people to access interventions anywhere, reduce costs, and potentially reach people who would not seek traditional treatment. Effectiveness varies widely across different apps and programs, with many lacking rigorous evaluation. But well-designed digital interventions show promise for extending access to evidence-based care.

Wearable devices and smartphone sensors can track sleep, activity, heart rate variability, and other markers potentially relevant to mental health. This passive data collection could allow early detection of symptom changes, triggering interventions before full relapse. However, concerns about privacy, accuracy, and who has access to data require careful consideration.

Neuromodulation techniques including transcranial magnetic stimulation and transcranial direct current stimulation are being explored for diverse applications beyond depression treatment including cognitive enhancement, stroke rehabilitation, and addiction treatment. While evidence supports some applications, enthusiasm has sometimes outpaced evidence, and optimal protocols remain under investigation.

Lifestyle interventions based on neuroscience principles including exercise programs, sleep interventions, stress reduction training, and dietary changes provide low-risk approaches to supporting brain health. Evidence increasingly shows that what is good for cardiovascular health is good for brain health, with exercise, healthy diet, social connection, cognitive engagement, and stress management all supporting cognitive function and potentially reducing dementia risk.

Educational applications of neuroscience include teaching methods aligned with how the brain learns, timing education to leverage sensitive periods, and accommodations for students with different learning profiles. While "brain-based learning" has sometimes been overhyped and neuromyths have proliferated, genuine insights from neuroscience can inform education when applied carefully and combined with educational research and practice.

Synthesis: Biology, Experience, and Human Nature

Throughout this exploration of biological psychology, several themes have emerged that illuminate the relationship between biology and experience in creating who we are.

The brain is both stable and plastic. Genes, early development, and learning create lasting brain features that shape personality, abilities, and vulnerabilities. Yet the brain remains capable of change throughout life, reorganizing in response to experience, learning, and intervention. This combination of stability and plasticity means we have consistent characteristics while retaining capacity for growth.

Biology and environment are inseparable. Every aspect of psychology reflects both biological and environmental influences working together. Genes require environments to be expressed. Environments affect biology through plasticity, epigenetics, and selection of environments. Attempting to partition variation into genetic versus environmental components, while useful for some purposes, obscures the reality that nature and nurture are inextricably intertwined.

Multiple levels of analysis are necessary. Understanding psychology requires examining genes, molecules, neurons, circuits, whole brains, behaviors, social contexts, and cultures. No single level is more "real" or fundamental than others. A complete account requires integrating across levels, understanding how processes at different scales interact to produce psychological phenomena.

Mechanisms matter for intervention. Descriptive knowledge of what symptoms co-occur or what factors predict outcomes is useful but limited. Understanding mechanisms, how symptoms arise from biological and psychological processes, enables more targeted and effective interventions. Biological psychology's focus on mechanisms provides leverage for developing better treatments.

Common mechanisms underlie diverse outcomes. The same brain systems and neurotransmitters influence multiple psychological functions. Stress response systems affect mood, cognition, immunity, and physical health. Dopamine influences motivation, pleasure, learning, and movement. This means interventions targeting common mechanisms may have broad benefits, but also that problems in one domain often cascade to affect others.

Individual differences are real and important. People differ in cognitive abilities, personality, emotional reactivity, and vulnerabilities to mental illness, reflecting both biological and environmental variation. Recognizing these differences allows tailoring education, career guidance, and treatment to individual characteristics. However, differences do not imply fixed limits, and most variation is continuous rather than creating discrete types.

Mental illness is biological but not simply biological. Depression, anxiety, schizophrenia, and other conditions involve measurable brain abnormalities, validating them as genuine medical conditions. But they also involve psychological, social, and cultural factors, requiring multifaceted treatment addressing biology, psychology, and environment.

We are more than our biology. While psychological phenomena have biological foundations, reducing human experience to biology alone misses emergent properties, meaning, values, and social contexts that shape lives. Biological psychology provides crucial insights but does not provide complete accounts of human nature.

Looking Forward: Challenges and Opportunities

The coming decades will likely bring continued rapid advances in biological psychology driven by new technologies, larger datasets, and better theories. Several challenges and opportunities will shape the field's trajectory.

Integrating big data from genomics, imaging, electronic health records, and other sources requires sophisticated computational approaches and raises questions about privacy, consent, and data sharing. Large consortia pooling data across sites enable studies with statistical power to detect small effects, but standardization of methods and protection of participant confidentiality are essential.

Translating discoveries into clinical applications remains challenging. Many biological findings in psychiatry have not yet led to improved treatments. The gap between understanding biological mechanisms and developing effective interventions reflects the complexity of mental illness and the difficulty of targeting specific mechanisms without affecting other systems. Continued investment in translational research is needed.

Addressing health disparities requires ensuring that advances in biological psychology benefit all people, not just privileged populations. Genetic studies have predominantly included European ancestry participants, limiting generalizability. Access to new treatments may be limited by costs. Cultural factors affect how mental illness is expressed and whether people seek help. Addressing these disparities requires inclusive research, affordable interventions, and cultural sensitivity.

Public communication of neuroscience findings must balance enthusiasm about discoveries with appropriate caveats about limitations and uncertainties. Neuromyths, misconceptions about the brain that persist despite being contradicted by evidence, proliferate partly due to oversimplified science communication. Responsible communication requires presenting findings accurately, acknowledging limitations, and resisting hype.

Interdisciplinary collaboration between neuroscience, psychology, psychiatry, genetics, computer science, philosophy, and other fields is essential for progress. Complex questions about brain and behavior require expertise from multiple disciplines. Training future scientists to work across disciplinary boundaries and creating incentives for collaboration will advance the field.

Global mental health requires extending evidence-based interventions to low-resource settings where most people with mental illness live. This requires not only making interventions affordable and accessible but also adapting them to different cultural contexts and addressing social determinants of mental health including poverty, violence, and discrimination.

Conclusion: The Promise of Biological Psychology

The journey through biological psychology reveals the brain as an organ of extraordinary complexity and sophistication, one that makes us who we are while remaining capable of change and adaptation. Understanding the biological foundations of psychology has already transformed mental health care, reduced stigma, and provided tools for enhancing human capacities. Yet this is just the beginning.

The coming decades promise continued discoveries about how the brain creates mind, how genes and environment interact to shape development, and how interventions can promote optimal function and treat dysfunction. These advances will raise new ethical challenges requiring thoughtful societal responses. They will create new possibilities for human enhancement and new risks of misuse. They will challenge existing concepts of human nature, responsibility, and identity.

Navigating this future wisely requires grounding applications of biological psychology in ethical principles including beneficence, non-maleficence, autonomy, and justice. It requires ensuring that knowledge is used to reduce suffering and promote flourishing rather than to control, discriminate, or exploit. It requires maintaining humility about what we know and acknowledging what remains uncertain. It requires respecting human dignity while pursuing scientific understanding.

For individuals, understanding biological psychology provides practical knowledge for protecting brain health, optimizing cognitive function, managing stress and emotions, and making informed decisions about mental health treatment. It offers perspective on the interplay of genes and environment, stability and change, strengths and vulnerabilities that make each person unique.

For society, biological psychology provides foundations for evidence-based policies in education, mental health care, criminal justice, and other domains affecting millions of lives. It offers hope for preventing and treating conditions that cause immense suffering. It reveals our shared biological heritage while illuminating the sources of human diversity.

The brain that has been the subject of this book is also its author and its reader. The same neural mechanisms described in these chapters allowed them to be written and allow you to comprehend them. This reflexive quality, the brain studying itself, gives neuroscience a unique character among sciences. We are simultaneously the observers and the observed, the scientists and the subjects, using our brains to understand how brains work.

This endeavor reveals both the power and limits of human understanding. We have made remarkable progress in understanding the biological basis of psychology, yet mysteries remain. The hard problem of consciousness persists. The precise mechanisms of memory storage are incompletely known. The origins of mental illness remain partly mysterious. These unknowns do not diminish what has been learned but rather highlight frontiers for future exploration.

The story of biological psychology is ultimately a story about what makes us human—our capacity for thought and emotion, learning and memory, connection and compassion, and also our vulnerability to suffering when these systems malfunction. It is a story that continues unfolding, with each generation of scientists adding new chapters. Understanding the biological foundations of psychology does not diminish the wonder of human experience but rather deepens appreciation for the remarkable biological processes that create it.

As we conclude this exploration, the central message is one of possibility and responsibility. We have unprecedented understanding of the brain and growing capacity to measure and modify its function. This knowledge empowers us to reduce suffering, enhance capacities, and promote flourishing.

But it also requires wisdom in application, compassion in implementation, and constant attention to ethical implications. The future of biological psychology lies not just in what we discover but in how we use that knowledge to create a world where all people can develop their potential and live fulfilling lives.

The brain you have been reading about is your brain, remarkably similar to all human brains yet uniquely shaped by your experiences. It has changed while reading this book, forming new connections, strengthening some pathways while allowing others to fade. It will continue changing throughout your life, plastic and adaptive, vulnerable yet resilient. Understanding it better is the first step toward caring for it wisely, using it well, and appreciating the extraordinary gift of consciousness it provides.

15

EPILOGUE: THE BRAIN IN CONTEXT

We have journeyed through the biological foundations of human psychology, exploring the structures, systems, and mechanisms that create mental life. From the intricate architecture of neurons to the complex circuits supporting memory and emotion, from the neurochemistry of mood to the plasticity enabling learning, from the genetic influences on development to the environmental factors that shape outcomes, we have seen how biology creates the substrate for everything we experience as psychological.

Yet as we reach the end of this exploration, it is worth stepping back to consider the broader context and implications of this biological understanding. What does it mean to know that our thoughts, feelings, and behaviors emerge from physical processes in the brain? How should this knowledge shape how we live, how we treat others, and how we organize society? What remains mysterious despite all we have learned, and what frontiers await future explorers of the brain?

The biological perspective on psychology is powerful but not complete. Yes, every mental process has a biological basis. Neural activity creates consciousness, neurotransmitters influence mood, genes affect personality, and brain structure constrains abilities. But acknowledging these biological foundations does not reduce human experience to biology alone. Multiple levels of explanation remain valid and necessary.

Consider love, explored in Chapter 8. We now know that romantic attachment involves specific neural circuits, that oxytocin promotes bonding, that dopamine creates the euphoria of new love, and that long-term relationships show different patterns of brain activation than new ones. This biological understanding is real and important. It explains why love feels the way it does, why heartbreak is so painful, and why some relationships form while others do not.

Yet describing the neurobiology of love does not capture its full meaning. Love exists simultaneously as biological process, psychological experience, social relationship, and cultural construction. The neural mechanisms enable love but do not exhaust its significance. Understanding oxytocin release does not make a marriage proposal less meaningful. Knowing that the amygdala processes emotional attachment does not diminish the joy of reuniting with a loved one after separation.

This multilevel reality applies throughout psychology. Depression involves altered brain chemistry and disrupted neural circuits, as we explored in Chapter 13, but it also involves negative thought patterns, stressful life circumstances, social isolation, and possibly spiritual crisis. Effective treatment addresses multiple levels simultaneously—medication to alter neurochemistry, psychotherapy to change thought patterns, social interventions to reduce isolation, and perhaps spiritual practices for those who find them meaningful. No single level is the "real" explanation with others being merely epiphenomenal.

The reductionist impulse to explain higher levels in terms of lower levels is scientifically productive but philosophically limited. Yes, psychological processes depend on biological processes, which depend on chemical processes, which depend on physical processes. But this dependence does not mean higher levels are illusory or unimportant. Emergent properties arise at each level that cannot be fully predicted from lower levels even though they depend on them. Consciousness emerges from neural activity but involves properties that purely neural descriptions miss. Social relationships emerge from individual psychologies but follow dynamics that individual-level descriptions cannot capture.

Understanding that mental life has biological foundations does not diminish human agency or responsibility. Some fear that if our thoughts and actions arise from prior brain states determined by genes and experience, then free will is illusory and responsibility disappears. But this confuses explanation with excuse. Explaining how decisions are made neurobiologically does not eliminate the reality that persons make choices that have consequences requiring accountability.

Moreover, the brain is not a deterministic machine executing fixed programs but rather a plastic organ continuously shaped by experience. The choices we make literally change our brains, creating feedback loops where biology affects choices and choices affect biology. This bidirectional causation preserves meaningful agency while acknowledging biological constraints. We are simultaneously shaped by our biology and shapers of it through our actions.

The practical implications of biological psychology extend across numerous domains. In mental healthcare, understanding the biological basis of psychiatric disorders has transformed treatment, reduced stigma, and improved outcomes. Recognizing depression as involving measurable brain abnormalities rather than character weakness encourages seeking help and validates suffering. Understanding anxiety disorders as involving overactive threat-detection systems suggests specific interventions targeting those systems. Appreciating that addiction hijacks reward circuits explains why willpower alone often fails and why comprehensive treatment addressing biological, psychological, and social factors works best.

In education, neuroscience insights about how the brain learns have begun influencing teaching methods, though translation has been uneven and neuromyths persist. Genuine insights include recognizing that memory consolidation requires sleep, that spacing learning over time beats cramming, that retrieval practice strengthens memory more than rereading, and that growth mindsets promoting belief in malleability improve motivation and outcomes. Less valid are claims about learning styles, right-brain/left-brain teaching, or critical periods that rigidly determine what can be learned when.

In criminal justice, understanding the biological basis of impulse control, decision-making, and antisocial behavior raises questions about culpability and punishment. Adolescents show different patterns of brain activation than adults when making risky decisions, with still-developing prefrontal cortex providing less regulatory control over limbic impulses. Should this neurobiological immaturity affect how we judge and sentence young offenders? Brain damage affecting regions supporting moral reasoning and impulse control sometimes precedes criminal behavior. Should this mitigate responsibility?

These questions have no easy answers and require balancing scientific understanding with legal principles and societal values. The point is not that neuroscience dictates answers but that biological understanding should inform deliberation about issues affecting millions of lives. Ignoring relevant scientific evidence produces sub optimal policies, while mechanically applying neuroscience findings without considering broader contexts and values also fails.

In the workplace, understanding individual differences in cognitive abilities, learning styles, and stress responses could enable better matching of people to roles and more effective training and development. Recognizing that attention has biological limits suggests designing work to respect those limits rather than expecting unlimited focus. Understanding that stress impairs decision-making suggests creating conditions that minimize unnecessary stress during critical decisions.

For individuals, biological psychology provides practical guidance for protecting brain health and optimizing function. The same factors promoting cardiovascular health—exercise, healthy diet, stress management, social connection, adequate sleep—also protect brain health and reduce dementia risk. Understanding that the brain remains plastic throughout life encourages continued learning and growth. Recognizing that mental health problems have biological components reduces self-blame while motivating evidence-based treatment.

Yet biological knowledge also creates risks. Genetic determinism the incorrect belief that genes rigidly determine outcomes could lead to giving up on people with genetic vulnerabilities. Neurobiological reductionism the belief that biological levels of explanation are more real than psychological or social levels could lead to ignoring important non-biological factors. Over medication of psychological distress could occur if biological interventions are pursued to the exclusion of psychological and social interventions. These risks require vigilance and balanced approaches integrating biological understanding with other forms of knowledge.

The ethical challenges raised by advancing neuroscience will intensify as technologies improve. Brain imaging is revealing aspects of mental life previously invisible. Genetic testing is identifying vulnerability factors before symptoms appear. Neurotechnologies are allowing direct modification of brain function. Artificial intelligence is decoding neural signals with increasing sophistication. Each advance creates new possibilities and new dilemmas requiring thoughtful societal responses.

Questions about privacy, autonomy, fairness, and human nature will arise with increasing urgency. Should genetic testing for mental illness vulnerability be encouraged, allowed, or restricted? Should cognitive enhancement be pursued, regulated, or prohibited? Should brain imaging be used in legal proceedings to assess credibility, predict dangerousness, or determine culpability? Should parents be allowed to select embryos based on genetic profiles predicting intelligence or personality? These questions do not have purely scientific answers but require ethical deliberation informed by scientific understanding.

As we have seen throughout this book, profound mysteries remain despite remarkable progress. The hard problem of consciousness why physical processes in the brain feel like something, why there is subjective experience at all remains unsolved. We can identify neural correlates of consciousness and manipulate brain states to alter consciousness, but explaining how and why neural activity creates the felt quality of experience eludes current science. This may be a question science eventually answers, or it may reflect fundamental limits of scientific explanation when the explained and explained are the same.

The precise mechanisms of memory storage remain incompletely understood. We know that memory involves changes in synaptic strengths and that specific brain regions support different memory types, but exactly how information is encoded in patterns of synaptic weights and how the brain retrieves specific memories from vast stores remain unclear. Understanding memory encoding at this detailed level would illuminate why memories sometimes fail and how to enhance memory function.

Individual differences in psychological characteristics reflect complex interactions between thousands of genetic variants and diverse environmental influences that we are only beginning to understand. Current genetic findings explain small fractions of heritability for most traits, suggesting much remains to be discovered. Disentangling genetic effects from environmental effects and understanding their interactions requires sophisticated methods and large datasets that are only now becoming available.

The development of mental illness involves interactions between genetic vulnerability, environmental stressors, and chance events that are incompletely predictable. Why some vulnerable people develop disorders while others remain healthy despite similar circumstances and vulnerabilities remains partly mysterious. Understanding these protective factors could enable interventions promoting resilience.

Despite these unknowns, the progress achieved over the past century provides grounds for optimism. The brain that once seemed hopelessly complex and forever beyond understanding has yielded many of its secrets to patient scientific investigation. Techniques that did not exist decades ago functional brain imaging, optogenetics, genome sequencing, machine learning now provide unprecedented insight. The pace of discovery is accelerating as new technologies enable questions that were previously unapproachable.

The ultimate goal of biological psychology is not merely understanding for its own sake, though the intellectual satisfaction of understanding how the brain works is considerable. The goal is using understanding to reduce suffering and promote flourishing. Every advance in understanding the biological basis of mental illness brings us closer to better treatments. Every insight into how the brain learns enables more effective education. Every discovery about factors promoting brain health empowers individuals to make better choices.

This knowledge is not the possession of experts alone but belongs to everyone. An educated citizenry understanding the biological basis of behavior can make better personal decisions about mental health treatment, education, lifestyle choices, and parenting. An informed public can engage more effectively in policy debates about mental health funding, educational reform, criminal justice, and research priorities. Democratic societies function best when important knowledge is widely shared rather than restricted to specialists.

As we conclude this exploration of the biology of the human mind, the central message is one of wonder, responsibility, and possibility. Wonder at the extraordinary complexity and sophistication of the biological processes creating conscious experience. Responsibility to use growing knowledge wisely and ethically, improving lives while avoiding misuse. Possibility for continued progress in understanding the brain, treating mental illness, enhancing education, and supporting human development.

The brain you have been reading about is your brain, similar to all human brains in basic organization yet uniquely shaped by your genetic inheritance and life experiences. It has been changing as you read, forming new connections, consolidating information into memory, and integrating new knowledge with what you already knew. It will continue changing throughout your life, plastic and adaptive, vulnerable yet remarkably resilient.

Understanding your brain better equips you to care for it wisely. Knowing that sleep consolidates memories motivates protecting sleep rather than sacrificing it for productivity. Understanding that chronic stress damages brain structure encourages stress management. Recognizing that social connection is biologically necessary rather than merely pleasant encourages maintaining relationships. Appreciating that the brain remains plastic throughout life inspires continued learning and growth.

For those who suffer from mental illness, understanding the biological foundations of their conditions can reduce shame and encourage seeking help. Depression, anxiety, addiction, and other disorders are not character flaws or moral failings but medical conditions involving biology gone awry. They deserve treatment as surely as diabetes or heart disease, and understanding them as brain disorders helps combat stigma.

For those who work in fields touching on human psychology educators, healthcare providers, managers, counselors, judges, policymakers biological understanding provides crucial context for their work. Recognizing that children's developing brains are particularly plastic suggests educational approaches leveraging this plasticity. Understanding that adolescent brains are still maturing explains behavior that might otherwise seem willfully irresponsible. Appreciating that stress impairs judgment suggests creating conditions that minimize stress during important decisions.

The study of biological psychology is far from complete. Each generation of researchers adds new chapters to the story, revealing previously hidden mechanisms, correcting earlier misunderstandings, and opening new questions. The students reading this book today may become the researchers making tomorrow's discoveries, using technologies we can barely imagine to answer questions we have not yet formulated clearly.

But even with much remaining unknown, what we have learned provides a sturdy foundation for understanding human nature. We are biological beings whose psychology emerges from brain processes shaped by genes and experience. We are simultaneously products of evolution's blind experimentation over millions of years and products of our unique developmental trajectories over individual lifetimes. We are constrained by our biology yet capable of transcending constraints through learning, culture, and technology. We are vulnerable to disorders when biological systems malfunction yet capable of remarkable recovery and resilience.

Most fundamentally, we are conscious beings whose awareness of our own nature sets us apart from other creatures. The brain has evolved to the point where it can study itself, understand itself, and modify itself. This reflexive capacity gives humans unique power and responsibility. We can choose, within limits, what kinds of minds we develop and what kinds of societies we build. Understanding the biological foundations of psychology is essential for making these choices wisely.

As you close this book and return to your life, you carry with you knowledge that connects the most intimate aspects of your experience with the physical world of biology. Your thoughts are neural activity. Your memories are synaptic strengths. Your emotions are coordinated patterns across brain regions. Your learning is plasticity. Your consciousness is the brain knowing itself.

This knowledge need not diminish the wonder of being human but rather deepens it. The fact that evolution and development have conspired to create beings capable of love, creativity, moral judgment, scientific understanding, and self-awareness from biological building blocks is extraordinary. The fact that we can understand, even partially, how this occurs is a testament to human ingenuity and persistence.

Use this knowledge well. Care for your brain through healthy lifestyle choices. Seek help when psychological struggles arise, knowing they reflect biology, not character. Approach others with compassion, recognizing that behavior reflects complex interactions between biology and experience. Support policies and practices informed by evidence about how brains actually work. Maintain wonder at the remarkable organ that makes you who you are.

The biology of the human mind is the story of how three pounds of tissue creates the richness of human experience. It is your story, the story of everyone you know, and the story of humanity. Understanding it better is the first step toward living more fully, thinking more clearly, and building a world where all brains can flourish.

The journey through biological psychology never truly ends. The brain continues changing, science continues discovering, and our understanding continues deepening. But this book provides a foundation a map of the territory, a framework for continued exploration, and an invitation to wonder at the extraordinary biological organ that makes us human.

Thank you for reading. May your brain serve you well.

THE END

www.ingramcontent.com/pod-product-compliance
Lightning Source LLC
LaVergne TN
LVHW052246200926
848481LV00002B/294
* 9 7 9 8 2 3 3 6 4 2 6 7 8 *